AF508672

Plant Embryogenesis

Plant Embryogenesis

Editors

Minako Ueda
Daisuke Kurihara

MDPI • Basel • Beijing • Wuhan • Barcelona • Belgrade • Manchester • Tokyo • Cluj • Tianjin

Editors
Minako Ueda
Tohoku University
Japan

Daisuke Kurihara
Nagoya University
Japan

Editorial Office
MDPI
St. Alban-Anlage 66
4052 Basel, Switzerland

This is a reprint of articles from the Special Issue published online in the open access journal *Plants* (ISSN 2223-7747) (available at: https://www.mdpi.com/journal/plants/special_issues/plant_embryogenesis).

For citation purposes, cite each article independently as indicated on the article page online and as indicated below:

LastName, A.A.; LastName, B.B.; LastName, C.C. Article Title. *Journal Name* **Year**, *Volume Number*, Page Range.

ISBN 978-3-0365-1461-1 (Hbk)
ISBN 978-3-0365-1462-8 (PDF)

Contents

About the Editors . . . . **vii**

Preface to "Plant Embryogenesis" . . . . **ix**

Hiroyuki Iida and Shinobu Takada
A Quarter Century History of *ATML1* Gene Research
Reprinted from: *Plants* **2021**, *10*, 290, doi:10.3390/plants10020290 . . . . **1**

**Ana D. Simonović, Milana M. Trifunović-Momčilov, Biljana K. Filipović, Marija P. Marković,
Milica D. Bogdanović and Angelina R. Subotić**
Somatic Embryogenesis in *Centaurium erythraea* Rafn—Current Status and Perspectives: A
Review
Reprinted from: *Plants* **2020**, *10*, 70, doi:10.3390/plants10010070 . . . . **9**

**Ayame Imoto, Mizuki Yamada, Takumi Sakamoto, Airi Okuyama, Takashi Ishida,
Shinichiro Sawa and Mitsuhiro Aida**
A ClearSee-Based Clearing Protocol for 3D Visualization of *Arabidopsis thaliana* Embryos
Reprinted from: *Plants* **2021**, *10*, 190, doi:10.3390/plants10020190 . . . . **29**

Ce Shi, Pan Luo, Peng Zhao and Meng-Xiang Sun
Detection of Embryonic Suspensor Cell Death by Whole-Mount TUNEL Assay in Tobacco
Reprinted from: *Plants* **2020**, *9*, 1196, doi:10.3390/plants9091196 . . . . **37**

Ryouya Deushi, Erika Toda, Shizuka Koshimizu, Kentaro Yano and Takashi Okamoto
Effect of Paternal Genome Excess on the Developmental and Gene Expression Profiles of
Polyspermic Zygotes in Rice
Reprinted from: *Plants* **2021**, *10*, 255, doi:10.3390/plants10020255 . . . . **45**

**Lourdes Delgado-Aceves, María Teresa González-Arnao, Fernando Santacruz-Ruvalcaba,
Raquel Folgado and Liberato Portillo**
Indirect Somatic Embryogenesis and Cryopreservation of *Agave tequilana* Weber Cultivar
'Chato'
Reprinted from: *Plants* **2021**, *10*, 249, doi:10.3390/plants10020249 . . . . **59**

**Milica D. Bogdanović, Katarina B. Ćuković, Angelina R. Subotić, Milan B. Dragićević,
Ana D. Simonović, Biljana K. Filipović and Slađana I. Todorović**
Secondary Somatic Embryogenesis in *Centaurium erythraea* Rafn
Reprinted from: *Plants* **2021**, *10*, 199, doi:10.3390/plants10020199 . . . . **73**

Xi Wei, Yanpeng Ding, Ye Wang, Fuguang Li and Xiaoyang Ge
Early Low-Fluence Red Light or Darkness Modulates the Shoot Regeneration Capacity of
Excised *Arabidopsis* Roots
Reprinted from: *Plants* **2020**, *9*, 1378, doi:10.3390/plants9101378 . . . . **93**

About the Editors

Minako Ueda

Apr, 2002–Mar, 2005 Graduate School of Science, Kyoto University, Department of Botany, JSPS fellow (DC1)

Apr, 2005–Mar, 2007 Institute of Biology III, University of Freiburg, JSPS fellow

Apr, 2007–Mar, 2008 Institute of Biology III, University of Freiburg, Postdoctoral fellow

Apr, 2008–Sep, 2010 Graduate School of Science, Nagoya University, Division of Biological Science, JSPS fellow (PD)

Oct, 2010–Mar, 2013 Nara Institute of Science and Technology

Apr, 2013–Sep, 2020 Nagoya University Institute of Transformative Bio-Molecules

Oct, 2020–Present Tohoku University Graduate School of Life Sciences, Department of Ecological Developmental Adaptability Life Sciences

Daisuke Kurihara

Apr, 2006–Mar, 2009 JSPS fellow (DC1)

Apr, 2009–Dec, 2010 JSPS fellow (PD)

Jan, 2011–Mar, 2017 Nagoya University Graduate School of Science, Designated Assistant Professor

Jan, 2011–Mar, 2017 JST ERATO Higashiyama Live-Holonics Project, Optical Technology Group, Group Leader

Apr, 2017–Sep, 2018 Nagoya University Graduate School of Science, Designated Lecturer

Oct, 2018–Mar, 2019 Nagoya University Institute of Transformative Bio-Molecules, Visiting Scholar

Oct, 2018–Present JST PRESTO Researcher

Apr, 2019–Present Nagoya University Institute of Transformative Bio-Molecules, Designated Lecturer

Preface to "Plant Embryogenesis"

Embryogenesis is a fundamental process in plant ontogeny. The fusion of a paternal sperm and a maternal egg generates a zygote and initiates the series of developmental events to set the basic body plan of the future plant. Therefore, embryogenesis is a dynamic procedure, involving a shift from the haploid gametophytic to the diploid sporophytic generation, metabolic activation, pattern formation, and dormancy in seed maturation. Furthermore, successful embryogenesis is essential for plant fertility and reproductive fitness. Thus, embryonic regulations are important not only for understanding both plant evolution and the diverse survival strategies of various plant species but also for bioengineering to increase plant productivity in agriculture.

Minako Ueda, Daisuke Kurihara
Editors

Review

A Quarter Century History of *ATML1* Gene Research

Hiroyuki Iida [1] and Shinobu Takada [2,*]

[1]   Institute of Biotechnology, HiLIFE, University of Helsinki, Viikinkaari 1, 00014 Helsinki, Finland; hiroyuki.iida@helsinki.fi

[2]   Department of Biological Sciences, Graduate School of Science, Osaka University, 1–1 Machikaneyama, Toyonaka, Osaka 560-0043, Japan

*   Correspondence: shinobu_takada@bio.sci.osaka-u.ac.jp

Abstract: The cloning of the *ATML1* gene, encoding an HD-ZIP class IV transcription factor, was first reported in 1996. Because *ATML1* mRNA was preferentially detected in the shoot epidermis, cis-regulatory sequences of *ATML1* have been used to drive gene expression in the outermost cells of the shoot apical meristem and leaves, even before the function of *ATML1* was understood. Later studies revealed that *ATML1* is required for developmental processes related to shoot epidermal specification and differentiation. Consistent with its central role in epidermal development, *ATML1* activity has been revealed to be restricted to the outermost cells via several regulatory mechanisms. In this review, we look back on the history of *ATML1* research and provide a perspective for future studies.

Keywords: embryogenesis; epidermal development; *ATML1*; transcriptional regulation; post-transcriptional regulation

Citation: Iida, H.; Takada, S. A Quarter Century History of *ATML1* Gene Research. *Plants* **2021**, *10* , 290 . https://doi.org/10.3390/plants10020290

Academic Editor: Minako Ueda

Received: 5 January 2021

Accepted: 29 January 2021

Published: 3 February 2021

1. Epidermis Formation and Its Role in Plant Development

ATML1 was first reported ~25 years ago as a gene that is expressed in the epidermis of *Arabidopsis thaliana*. The epidermis is the outermost cell layer of land plants, first formed during early embryogenesis. In *Arabidopsis thaliana*, embryo proper cells during the eight-cell stage undergo tangential cell divisions, generating outer and inner daughter cells. The outer daughter cells exhibit mostly anticlinal cell divisions to help maintain the single cell layer and are eventually differentiated into the epidermal cells.

The epidermis is positioned at the interface between plants and the environment; therefore, specialized cell types and structures in the epidermis facilitate abiotic and biotic stress responses. For example, stomatal guard cells and trichomes enable efficient gas exchange and prevent insect herbivory, respectively [1,2]. Additionally, the cuticle, a hydrophobic lipid layer, is deposited on the outermost surface of the shoot epidermal cell walls, in order to prevent water loss due to the dry terrestrial environment [3].

In addition to its protective functions, the epidermis plays an important role in controlling plant growth. Brassinosteroid (BR) is a plant hormone promoting cell division and expansion and the BR response and biosynthetic mutants show a severe dwarf phenotype. Activation of BR signaling in the outermost layer of those mutants rescued their dwarf phenotypes, whereas inactivation of BR signaling in the wild-type epidermal layer made plants small. These observations suggest that the epidermal layer restricts the extent of shoot growth depending on the activity of BR signaling [4]. It has also been reported that production of very-long-chain fatty acids (VLCFAs) in the epidermis, which are components of the cuticle wax, is required for repressing cell proliferation in the vascular tissue [5,6]. Observed increase in cytokinin, a plant hormone that promotes cell proliferation, in VLCFA-deficient seedlings suggests that VLCFAs or their derivatives play a role in repressing cytokinin biosynthesis in the vasculature [5].

The epidermis also influences pattern formation in the shoot apical meristem (SAM), as several reports have suggested that certain microRNAs (miRNAs), generated in the

Plants **2021**, *10* , 290 . https://doi.org/10.3390/plants10020290   https://www.mdpi.com/journal/plants

outermost cell layer of the SAM, form an inhibitory gradient that contributes to the positioning of the stem cell niche at away from the epidermis [7,8]. Consistently, epidermal deficient mutants often show ectopic SAM activity in the leaves [9].

Molecular genetic analysis has revealed many genes that are required for epidermal cell differentiation. Among those genes, *ARABIDOPSIS THALIANA MERISTEM L1 LAYER* (*ATML1*), which encodes an HD-ZIP class IV transcription factor, has been shown to be a master regulator for shoot epidermal cell identity in *Arabidopsis thaliana* (Figure 1). This review summarizes *ATML1* research of the last 25 years and provides a future perspective in the field.

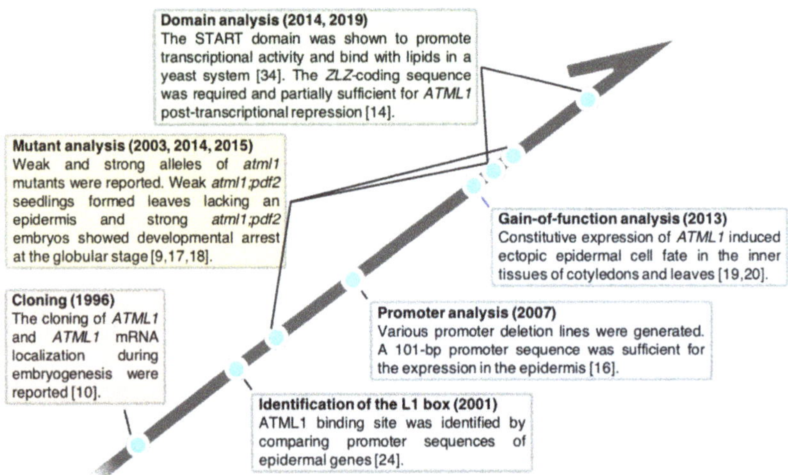

Figure 1. Timeline of the *ATML1* research. The history of *ATML1* gene research is summarized. Each box shows important findings in the indicated year.

2. Cloning of the *ATML1* Gene and Analysis of Its Expression Pattern

ATML1 was first identified in a screen searching for cDNA clones that have a sequence homology with an ovule-enriched homeobox gene, O39 in orchid, by plaque hybridization [10,11]. 'ATML1' was initially named as an acronym of <u>A</u>RABIDOPSIS <u>TH</u>ALIANA <u>M</u>ERISTEM <u>L1</u> LAYER (the outermost layer is called L1) although it is mistakenly written down as *AtML1* or has been described with the acronym of *ARABIDOPSIS THALIANA MERISTEM LAYER1*. *ATML1* belongs to the HD-ZIP class IV family transcription factors, including 16 members in *Arabidopsis thaliana*. The first identified member of the HD-ZIP class IV genes, *GLABRA2* (*GL2*), was involved in trichome and root hair patterning [12,13]. The first report showed that *ATML1* mRNA was detected specifically in the outermost cells of embryos, the SAM, and floral organs; *ATML1* is the first gene shown to be expressed in a specific cell layer of plant tissues [10] (Figure 2a,c).

In the first report, in situ hybridization experiments showed preferential accumulation of *ATML1* mRNA in the outermost cells of the SAM, leaf primordia, and floral organs. During embryogenesis, *ATML1* mRNA was detected in the apical cell of the one-cell stage embryo and its localization was restricted to the outermost cells of the embryo-proper from the early globular stage onwards [10].

After the first report, *ATML1* expression pattern has been re-examined in more details by several researchers [14–16]. *ATML1* promoter activity was also detected in the outermost cells of the root meristem [15] (Figure 2c). Detailed expression analysis in the embryos, with the aid of a sensitive reporter gene, revealed that *ATML1* promoter was also active in the basal cell of the one-cell stage embryo and in the suspensor cells (Figure 2a). Furthermore, when the reporter was fused with a destruction-box motif, which degrades the reporter protein at anaphase and enables detection of only newly synthesized protein after cell division, *ATML1* promoter activity was hardly detected in the inner cells of the 16-cell stage embryos, suggesting that *ATML1* transcription is restricted to the outermost cells from the 16-cell stage [16]. Recently, we reported that although *ATML1* was strongly expressed in the outermost cells, weak expression was also detected in the inner cells of the embryos, especially in the subepidermal cells [14] (Figure 2a).

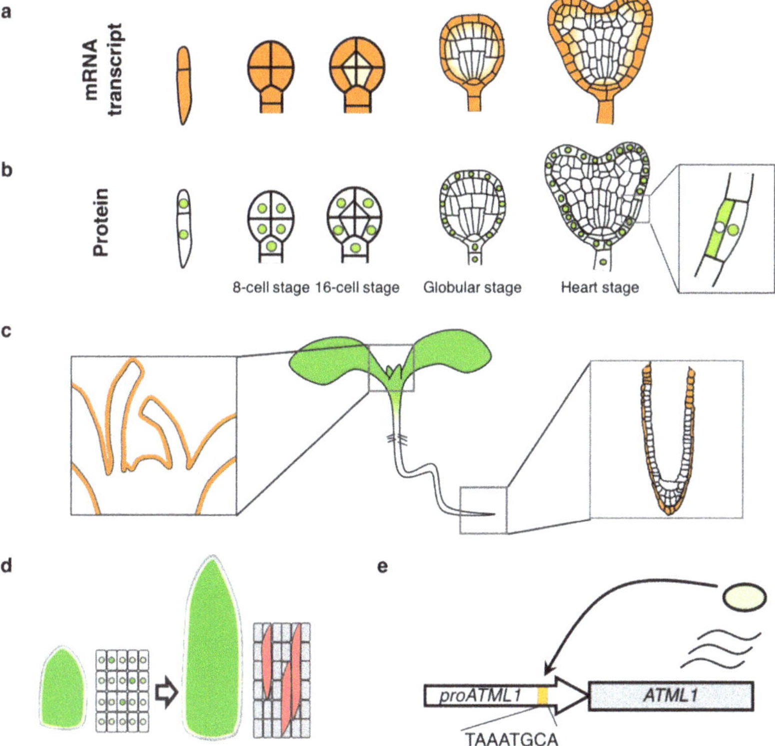

Figure 2. Schematic of *ATML1* transcript and ATML1 protein localization. (**a**) *ATML1* is transcribed strongly in the outermost cells and weakly in the inner cells of the embryos (**b**) ATML1 protein accumulation is observed only in the surface cells from the 16-cell stage. ATML1 protein is exclusively localized in the nuclei until the 8-cell stage. From the 16-cell stage onwards, ATML1 protein is present in both nuclei and cytosol of the outermost cells of the embryo-proper, while it only remains to accumulate in the nuclei of the suspensor cells. When the epidermal cells undergo aberrant periclinal divisions, ATML1 nuclear accumulation becomes weaker in the inner daughter cells compared with the outer daughter cells. (**c**) *ATML1* activity during post-embryonic development in the shoot apex (left) and the root tip (right). (**d**) When accumulated above the threshold level during the G2 phase, ATML1 promotes giant cell formation in the sepal epidermis. (**e**) Feedback regulation of *ATML1* via the L1 box in the promoter (*proATML1*). Orange: *ATML1* transcription, green in the cell: ATML1 protein, red: giant cells, yellow box in the *ATML1* promoter: an L1 box.

In situ localization of *ATML1* transcripts reported in 1996 implied its role in epidermis development, which was later proved in 2003.

3. *ATML1* Functions in Plant Development

ATML1 loss-of-function mutant phenotypes were first described in 2003 and its role in plant development was explored. The *atml1-1* single mutant, which has a T-DNA insertion near the 3′ end of the ORF, did not show a clear abnormality [17]. *PROTODERMAL FACTOR2* (*PDF2*), an *ATML1* paralog expressed in the epidermis, functions redundantly with *ATML1*; the *atml1-1;pdf2-1* double mutant is seedling-lethal harboring reduced cotyledons and narrow leaves lacking an epidermis [17]. Transcript levels of some epidermal genes were reduced in the *atml1-1;pdf2-1* mutant background, suggesting that *ATML1* and *PDF2* are required for epidermal differentiation. After 11 years, *atml1-1* was shown to be a weak allele mutant and a new *ATML1* null mutation, caused by a T-DNA insertion in the homeobox, combined with a *PDF2* mutation turned out to cause developmental arrest at the globular stage. The outermost cells of the arrested *atml1;pdf2* embryos were swollen and often underwent unusual periclinal cell divisions, suggesting that epidermal cell identity was lost in those embryos [9,18]. These results revealed that *ATML1* and *PDF2* are required for proper embryo development as well as epidermal specification. Considering that the epidermis is the first tissue formed during embryogenesis, the epidermis formation might be a prerequisite for the proper progression of embryogenesis.

Gain-of-function experiments have also been performed to assess the ability of *ATML1* to promote epidermal cell differentiation. When *ATML1* expression was constitutively induced in whole seedlings, epidermal traits such as stomatal guard cells and trichomes were found ectopically in the inner tissues of cotyledons or leaves [19,20]. Given that ectopic expression of a guard-cell marker was detected in *ATML1*-overexpressing roots, ATML1 should promote shoot but not root epidermal cell identity [20]. Furthermore, expression of *ATML1* and *PDF2* fused to a transcriptional repressor sequence (called EAR or SRDX motif) caused organ fusions and high permeability to a hydrophobic dye, which are typical phenotypes related to cuticle formation defects [9,21]. Expression of *ATML1-SRDX*, which is supposed to downregulate target genes of ATML1, caused de-differentiation of the epidermis even at late stages of seedling development, suggesting that ATML1 and/or its target genes are required for epidermal cell fate maintenance [21]. Additionally, *ATML1* promotes the differentiation of giant cells during flower development; in *Arabidopsis thaliana*, sepal epidermal cells come in different sizes: large 'giant cells' and small cells. The *atml1-2* mutation lacked giant cell formation, while *ATML1* overexpression resulted in sepals consisting of mostly giant cells [22,23]. Fluctuations of ATML1 protein levels occurred in sepal epidermal cells and when its level was above a certain threshold during the G2 phase, these cells would undergo endoreduplication and became giant cells [22] (Figure 2d). Therefore, *ATML1* is a key regulator not only for epidermal specification during embryogenesis but also for maintaining and specializing epidermal cells even at later development stages.

4. Transcriptional Regulation of the *ATML1* Gene

Multiple studies address how *ATML1* transcription is preferentially activated in the outermost cells, for example, by analyzing cis-regulatory sequences of *ATML1* [15,16,24]. Sequence comparison of L1-specific promoters, including *ATML1*, has identified a conserved cis-regulatory sequence, called L1 box. Direct binding of ATML1 and PDF2 to the L1 box has been shown in vitro and mutations in the L1 box abolished the expression of an epidermal gene, *PDF1* [17,24]. This has provided a molecular basis for epidermal gene activation by ATML1: *ATML1* activates epidermis-specific gene expression through its direct binding to the L1 box. Consistently, expression of epidermal genes that contain the L1 box in their regulatory regions was increased and decreased in *ATML1*-overexpressing and *atml1;pdf2* mutant plants, respectively [20]. Binding of ATML1 to the L1 box region was confirmed also in vivo by chromatin immunoprecipitation assays [18]. Because the *ATML1* promoter contains an L1 box sequence, it has been proposed that positive feedback regulation may be important for epidermis-specific expression of *ATML1* [24] (Figure 2e).

The whole genomic fragment of *ATML1* and the 3.4-kb promoter sequence, which has been widely used as a driver for epidermis-specific expression, drove the same expression pattern, suggesting that the 3.4-kb promoter sequence contains all the required cis-elements for proper *ATML1* expression [14,16]. For more precise understanding of

ATML1 transcriptional regulation, effects of deletions in the 3.4-kb *ATML1* promoter on the outermost-cell specific expression were examined in the embryos. Detailed promoter analysis has shown that a 101-bp *ATML1* promoter fragment, located upstream of the translational start site and contains an L1 box and a WUS-binding site, was sufficient for the expression in the outermost cells. Unexpectedly, even when mutating both the L1 box and WUS-binding site, reporter gene expression was reduced but still detected in the outermost cells, indicating that the expression in the epidermal cells can be achieved independently of these cis-elements. Consistently, *ATML1* expression was still detected in the outermost cells of *atml1-1;pdf2-1* embryos and leaves, suggesting that positive feedback regulation is not required for the outermost-cell specific promoter activity [16,20].

There are two scenarios that can explain the outermost cell-specific expression of *ATML1*: transcriptional activation in the outermost cells or transcriptional repression in the inner cells. Although it is difficult to distinguish these two possibilities, reporter gene expression was not activated in the inner cells of any promoter-deletion lines tested, imply-ing that there is no transcriptional repressors that directly bind to the *ATML1* promoter and repress its expression in the inner cells of the embryos [16]. Therefore, only positive regulators appear to act on the *ATML1* promoter to activate expression in the outermost cells, suggesting that transcriptional activation in the outermost cells is more plausible. However, we cannot exclude the possibility that these positive regulators are repressed or inactivated in the inner cells of the embryos.

What could be the candidate factors that activate *ATML1* transcription in the outer cells? First, full epidermal cell identity seems not to be required for *ATML1* expression because *ATML1* promoter activity was still detected in the outermost mesophyll-looking cells in the *atml1-1;pdf2-1* mutant [20]. Secondly, *ATML1* expression in the epidermis was detected in mutants defective in auxin signaling, suggesting that auxin, a plant hormone known to direct many cell fate decisions, is not involved in the transcriptional control of *ATML1* [16]. Considering that cell fate decisions in plants largely depend on the position of the cell, *ATML1* expression could be influenced by the position of the cell; namely, the "surface" position [25]. In agreement with this idea, when the epidermal cells underwent aberrant periclinal divisions, only the outer daughter cells remained as epidermal cells while the inner daughter cells were differentiated into mesophyll cells [26]. In addition, *ATML1* was still transcribed in the outermost cells in a mutant defective in the control of cell division orientation (unpublished results from the authors). These results suggest that cell position but not cell lineage is required for *ATML1* transcription, as well for epidermal/mesophyll cell fate decisions. To date, positional cues required for epider-mal/mesophyll cell fate changes have still remained unknown, however. Several mutants, in which *ATML1* transcription was decreased or lost, have been reported, implying that the causal genes of these mutants are positive regulators of *ATML1* transcription [27–29]. However, *ATML1* transcription could still be detected at early stages in these mutant embryos, suggesting that these causal genes may be involved, either directly or indirectly, in the maintenance of *ATML1* expression rather than its initiation [3]. Characterization of mutants that completely abolish or alter the *ATML1* expression pattern or direct purifica-tion of *ATML1* transcription-promoting molecules should be needed for further progress of *ATML1* research.

5. Post-transcriptional and Post-translational Regulation of *ATML1*

Several studies have implied post-transcriptional regulation of *ATML1*. In 2010, muta-tions in the *DICER-LIKE1* (*DCL1*) gene, which is required for correct processing of miRNAs, were shown to increase *ATML1* mRNA levels in suspensor cells whereas normally *ATML1* mRNA is under the detection limit in the wild-type suspensor cells [10,30]. This implies that *ATML1* mRNA is degraded in suspensor cells by unidentified miRNAs. However, the epidermis-specific *ATML1* mRNA accumulation was not affected in *dcl1* embryos, suggesting that miRNAs are not used as positional cues for epidermal specification [30]. Moreover, *ATML1* transcription and ATML1 protein accumulation were detected in both embryo proper and suspensor cells of wild-type embryos at similar levels using sensitive reporter lines, implying that protein accumulation is not affected despite the apparent differences in mRNA levels [14,16].

The idea of *ATML1* regulation at the protein level came from domain analysis. ATML1 protein is characterized by a homeodomain (HD), a zipper-loop-zipper (ZLZ) motif, a StAR-related lipid-transfer (START) domain and a START-associated domain (SAD) [31]. The HD domain is a DNA binding domain and ZLZ motif is required for dimerization [32]. The START domain is homologous to a lipid-sterol binding domain [33]. The function of SAD domain has not been elucidated. The START domains from ATML1, PDF2 and GL2 were implied to bind with lipids in a yeast system. Although transcriptional activity increased upon its binding with lipids, the START domain alone was not sufficient to confer transcriptional activation, suggesting that it influences the transcription in combination with other domains or co-factors [34]. Therefore, it is plausible that ATML1 is activated by interaction with signaling molecules including lipids through its START domain.

Recently, we have reported post-transcriptional regulation of *ATML1* in the embryos. We found that ATML1 protein was not detected in the inner cells of the embryos whereas *ATML1* was weakly transcribed in these cells (Figure 2a,b). In addition, when epidermal cells underwent periclinal cell divisions, nuclear localization of ATML1 was weaker in the inner daughter cells compared with the outer daughter cells (Figure 2b). Constitutive expression of *ATML1* also confirmed that nuclear localization was reduced in the inner tissues of the embryos. These observations suggest that ATML1 protein accumulation and nuclear localization are negatively regulated in the inner cells to restrict *ATML1* activity to the outermost cells. Treatment with a nuclear export inhibitor or domain deletions increased protein accumulation and nuclear localization of ATML1 in both the outermost and inner cells of the embryos, suggesting that *ATML1* activity was negatively regulated also in the outermost cells [14]. Because *ATML1* can potentially activate and maintain its own expression through positive feedback, post-transcriptional repression might be a reasonable mechanism to suppress excessive *ATML1* activity.

Domain deletion analysis revealed that the *ZLZ*-encoding sequence is required and partially sufficient for the post-transcriptional regulation. How the dimerization motif controls ATML1 protein accumulation and subcellular localization is unclear. One possibility is that protein-protein interaction is required for degradation or for nuclear export of ATML1 protein. The START domain was not necessary for the post-transcriptional repression in the outermost cells but was required for the repression of nuclear localization in the inner cells [14]. The START domain may, therefore, play a role in negatively regulating ATML1 activity in the inner cells by physical interaction with other molecules present in the inner tissues. To understand the mechanisms regulating *ATML1* activity at the post-transcriptional level, future work should focus on the identification of ATML1 interacting factors. DELLA proteins, which are central repressors of gibberellin (GA) signaling, were shown to interact with ATML1 and PDF2, probably to enable GA-dependent ATML1/PDF2 activation during seed germination [35]. In cotton, Gossypium barbadense Meristem Layer 1 (GbML1), an *ATML1* ortholog, has been shown to interact with a MYB domain transcription factor, GbMYB25 [32]. However, as far as we know, interaction of ATML1 or its orthologous protein with molecules that reduce nuclear accumulation or stability has not been reported yet.

6. Future Perspective

ATML1 has become an indispensable tool and research target for elucidating epidermal differentiation. Despite being studied for more than two decades, several important questions remain to be answered. Firstly, molecular mechanisms that restrict *ATML1* activity in the outermost cells have not been identified. Molecules directly interacting with ATML1 protein or *ATML1* cis-regulatory sequences may convey "surface" positional cues. Secondly, although *ATML1* orthologues have been identified and shown to be preferentially expressed in the outermost cells in other plant species, loss-of-function mutant phenotypes and biological roles of those genes have yet to be examined [36–38]. From an evolutionary point of view, an *ATML1* ancestral gene should have acquired the function to promote the expression of genes required for epidermal features in the outermost cells during or before plant terrestrialization, as the epidermis is an essential tissue for land plants to live on dry lands. It would be interesting to identify the *ATML1* ancestral gene and study how its regulation has been evolved during the transition from aquatic to land plants. Further *ATML1* research would shed light on developmental and evolutionary processes of land plants.

Author Contributions: All authors wrote the manuscript and prepared Figures. All authors have read and agreed to the published version of the manuscript.

Funding: This work was supported by grants from the Japan Society for the Promotion of Science to S.T. (JP20657012, JP22687003, JP23657036, JP26440142 and JP18K06286).

Institutional Review Board Statement: Not applicable.

Informed Consent Statement: Not applicable.

Data Availability Statement: Not applicable.

Acknowledgments: We thank Brecht Wybouw (University of Helsinki) for helpful comments on this manuscript.

Conflicts of Interest: The authors declare no conflict of interest.

References

1. Pattanaik, S.; Patra, B.; Singh, S.K.; Yuan, L. An overview of the gene regulatory network controlling trichome development in the model plant, *Arabidopsis*. *Front. Plant Sci.* **2014**, *5*, 259. [CrossRef] [PubMed]
2. Pillitteri, L.J.; Torii, K.U. Mechanisms of stomatal development. *Annu. Rev. Plant Biol.* **2012**, *63*, 591–614. [CrossRef] [PubMed]
3. Takada, S.; Iida, H. Specification of epidermal cell fate in plant shoots. *Front. Plant Sci.* **2014**, *5*, 49. [CrossRef] [PubMed]
4. Savaldi-Goldstein, S.; Peto, C.; Chory, J. The epidermis both drives and restricts plant shoot growth. *Nature* **2007**, *446*, 199–202. [CrossRef] [PubMed]
5. Nobusawa, T.; Okushima, Y.; Nagata, N.; Kojima, M.; Sakakibara, H.; Umeda, M. Synthesis of very-long-chain fatty acids in the epidermis controls plant organ growth by restricting cell proliferation. *PLoS Biol.* **2013**, *11*, e1001531. [CrossRef] [PubMed]
6. Yephremov, A.; Wisman, E.; Huijser, P.; Huijser, C.; Wellesen, K.; Saedler, H. Characterization of the FIDDLEHEAD gene of *Arabidopsis* reveals a link between adhesion response and cell differentiation in the epidermis. *Plant Cell* **1999**, *11*, 2187–2201. [PubMed]
7. Han, H.; Yan, A.; Li, L.; Zhu, Y.; Feng, B.; Liu, X.; Zhou, Y. A signal cascade originated from epidermis defines apical-basal patterning of *Arabidopsis* shoot apical meristems. *Nat. Commun.* **2020**, *11*, 1–17.
8. Knauer, S.; Holt, A.L.; Rubio-Somoza, I.; Tucker, E.J.; Hinze, A.; Pisch, M.; Javelle, M.; Timmermans, M.C.; Tucker, M.R.; Laux, T. A protodermal miR394 signal defines a region of stem cell competence in the *Arabidopsis* shoot meristem. *Dev. Cell* **2013**, *24*, 125–132. [CrossRef]
9. Ogawa, E.; Yamada, Y.; Sezaki, N.; Kosaka, S.; Kondo, H.; Kamata, N.; Abe, M.; Komeda, Y.; Takahashi, T. ATML1 and PDF2 play a redundant and essential role in *Arabidopsis* embryo development. *Plant Cell Physiol.* **2015**, *56*, 1183–1192. [CrossRef]
10. Lu, P.; Porat, R.; Nadeau, J.A.; O'Neill, S.D. Identification of a meristem L1 layer-specific gene in *Arabidopsis* that is expressed during embryonic pattern formation and defines a new class of homeobox genes. *Plant Cell* **1996**, *8*, 2155–2168.
11. Nadeau, J.A.; Zhang, X.S.; Li, J.; O'Neill, S.D. Ovule development: Identification of stage-specific and tissue-specific cDNAs. *Plant Cell* **1996**, *8*, 213–239. [PubMed]
12. Rerie, W.G.; Feldmann, K.A.; Marks, M.D. The GLABRA2 gene encodes a homeo domain protein required for normal trichome development in *Arabidopsis*. *Genes Dev.* **1994**, *8*, 1388–1399. [CrossRef] [PubMed]
13. Masucci, J.D.; Rerie, W.G.; Foreman, D.R.; Zhang, M.; Galway, M.E.; Marks, M.D.; Schiefelbein, J.W. The homeobox gene GLABRA2 is required for position-dependent cell differentiation in the root epidermis of *Arabidopsis thaliana*. *Development* **1996**, *122*, 1253–1260. [PubMed]
14. Iida, H.; Yoshida, A.; Takada, S. ATML1 activity is restricted to the outermost cells of the embryo through post-transcriptional repressions. *Development* **2019**, *146*. [CrossRef] [PubMed]
15. Sessions, A.; Weigel, D.; Yanofsky, M.F. The *Arabidopsis thaliana* MERISTEM LAYER 1 promoter specifies epidermal expression in meristems and young primordia. *Plant J.* **1999**, *20*, 259–263. [CrossRef] [PubMed]
16. Takada, S.; Jürgens, G. Transcriptional regulation of epidermal cell fate in the *Arabidopsis* embryo. *Development* **2007**, *134*, 1141–1150. [CrossRef] [PubMed]
17. Abe, M.; Katsumata, H.; Komeda, Y.; Takahashi, T. Regulation of shoot epidermal cell differentiation by a pair of homeodomain proteins in *Arabidopsis*. *Development* **2003**, *130*, 635–643. [CrossRef]

18. San-Bento, R.; Farcot, E.; Galletti, R.; Creff, A.; Ingram, G. Epidermal identity is maintained by cell–cell communication via a universally active feedback loop in *Arabidopsis thaliana*. *Plant J.* **2014**, *77*, 46–58. [CrossRef]
19. Peterson, K.M.; Shyu, C.; Burr, C.A.; Horst, R.J.; Kanaoka, M.M.; Omae, M.; Sato, Y.; Torii, K.U. *Arabidopsis* homeodomain-leucine zipper IV proteins promote stomatal development and ectopically induce stomata beyond the epidermis. *Development* **2013**, *140*, 1924–1935. [CrossRef]
20. Takada, S.; Takada, N.; Yoshida, A. ATML1 promotes epidermal cell differentiation in *Arabidopsis* shoots. *Development* **2013**, *140*, 1919–1923. [CrossRef]
21. Takada, S. Post-Embryonic induction of ATML1-SRDX alters the morphology of seedlings. *PLoS ONE* **2013**, *8*, e79312. [CrossRef] [PubMed]
22. Meyer, H.M.; Teles, J.; Formosa-Jordan, P.; Refahi, Y.; San-Bento, R.; Ingram, G.; Jonsson, H.; Locke, J.C.W.; Roeder, A.H.K. Fluctuations of the transcription factor ATML1 generate the pattern of giant cells in the *Arabidopsis* sepal. *eLife* **2017**, *6*, e19131. [CrossRef] [PubMed]
23. Roeder, A.H.; Cunha, A.; Ohno, C.K.; Meyerowitz, E.M. Cell cycle regulates cell type in the *Arabidopsis* sepal. *Development* **2012**, *139*, 4416–4427. [CrossRef] [PubMed]
24. Abe, M.; Takahashi, T.; Komeda, Y. Identification of a cis-regulatory element for L1 layer-specific gene expression, which is targeted by an L1-specific homeodomain protein. *Plant J.* **2001**, *26*, 487–494. [CrossRef] [PubMed]
25. Scheres, B. Plant cell identity. The role of position and lineage. *Plant Physiol.* **2001**, *125*, 112–114. [CrossRef] [PubMed]
26. Stewart, R.N.; Dermen, H. Flexibility in ontogeny as shown by the contribution of the shoot apical layers to leaves of periclinal chimeras. *Am. J. Bot.* **1975**, *62*, 935–947. [CrossRef]
27. Johnson, K.L.; Degnan, K.A.; Ross Walker, J.; Ingram, G.C. AtDEK1 is essential for specification of embryonic epidermal cell fate. *Plant J.* **2005**, *44*, 114–127. [CrossRef]
28. Nodine, M.D.; Yadegari, R.; Tax, F.E. RPK1 and TOAD2 are two receptor-like kinases redundantly required for *Arabidopsis* embryonic pattern formation. *Dev. Cell* **2007**, *12*, 943–956. [CrossRef]
29. Tanaka, H.; Watanabe, M.; Sasabe, M.; Hiroe, T.; Tanaka, T.; Tsukaya, H.; Ikezaki, M.; Machida, C.; Machida, Y. Novel receptor-like kinase ALE2 controls shoot development by specifying epidermis in *Arabidopsis*. *Development* **2007**, *134*, 1643–1652. [CrossRef]
30. Nodine, M.D.; Bartel, D.P. MicroRNAs prevent precocious gene expression and enable pattern formation during plant embryogenesis. *Genes Dev.* **2010**, *24*, 2678–2692. [CrossRef]
31. Ariel, F.D.; Manavella, P.A.; Dezar, C.A.; Chan, R.L. The true story of the HD-Zip family. *Trends Plant Sci.* **2007**, *12*, 419–426. [CrossRef] [PubMed]
32. Zhang, F.; Zuo, K.; Zhang, J.; Liu, X.; Zhang, L.; Sun, X.; Tang, K. An L1 box binding protein, GbML1, interacts with GbMYB25 to control cotton fibre development. *J. Exp. Bot.* **2010**, *61*, 3599–3613. [CrossRef] [PubMed]
33. Schrick, K.; Nguyen, D.; Karlowski, W.M.; Mayer, K.F. START lipid/sterol-binding domains are amplified in plants and are predominantly associated with homeodomain transcription factors. *Genome Biol.* **2004**, *5*, R41. [CrossRef] [PubMed]
34. Schrick, K.; Bruno, M.; Khosla, A.; Cox, P.N.; Marlatt, S.A.; Roque, R.A.; Nguyen, H.C.; He, C.; Snyder, M.P.; Singh, D.; et al. Shared functions of plant and mammalian StAR-related lipid transfer (START) domains in modulating transcription factor activity. *BMC Biol.* **2014**, *12*, 70. [CrossRef] [PubMed]
35. Rombolá-Caldentey, B.; Rueda-Romero, P.; Iglesias-Fernández, R.; Carbonero, P.; Oñate-Sánchez, L. *Arabidopsis* DELLA and two HD-ZIP transcription factors regulate GA signaling in the epidermis through the L1 box cis-element. *Plant Cell* **2014**, *26*, 2905–2919. [CrossRef] [PubMed]
36. Ito, M.; Sentoku, N.; Nishimura, A.; Hong, S.-K.; Sato, Y.; Matsuoka, M. Position dependent expression of GL2-type homeobox gene, Roc1: Significance for protoderm differentiation and radial pattern formation in early rice embryogenesis. *Plant J.* **2002**, *29*, 497–507. [CrossRef] [PubMed]
37. Javelle, M.; Klein-Cosson, C.; Vernoud, V.; Boltz, V.; Maher, C.; Timmermans, M.; Depège-Fargeix, N.; Rogowsky, P.M. Genome-wide characterization of the HD-ZIP IV transcription factor family in maize: Preferential expression in the epidermis. *Plant Physiol.* **2011**, *157*, 790–803. [CrossRef]
38. Gao, L.; Tian, Y.; Chen, M.-C.; Wei, L.; Gao, T.-G.; Yin, H.-J.; Zhang, J.-L.; Kumar, T.; Liu, L.-B.; Wang, S.-M. Cloning and functional characterization of epidermis-specific promoter MtML1 from *Medicago truncatula*. *J. Biotechnol.* **2019**, *300*, 32–39. [CrossRef]

Review

Somatic Embryogenesis in *Centaurium erythraea* Rafn—Current Status and Perspectives: A Review

Ana D. Simonović, Milana M. Trifunović-Momčilov *, Biljana K. Filipović , Marija P. Marković, Milica D. Bogdanović and Angelina R. Subotić

Department for Plant Physiology, Institute for Biological Research "Siniša Stanković"—National Institute of Republic of Serbia, University of Belgrade, Bulevar Despota Stefana 142, 11060 Belgrade, Serbia; ana.simonovic@ibiss.bg.ac.rs (A.D.S.); biljana.nikolic@ibiss.bg.ac.rs (B.K.F.); marija.nikolic@ibiss.bg.ac.rs (M.P.M.); milica.bogdanovic@ibiss.bg.ac.rs (M.D.B.); heroina@ibiss.bg.ac.rs (A.R.S.)
* Correspondence: milanag@ibiss.bg.ac.rs

Abstract: *Centaurium erythraea* (centaury) is a traditionally used medicinal plant, with a spectrum of secondary metabolites with confirmed healing properties. Centaury is an emerging model in plant developmental biology due to its vigorous regenerative potential and great developmental plasticity when cultured in vitro. Hereby, we review nearly two decades of research on somatic embryogenesis (SE) in centaury. During SE, somatic cells are induced by suitable culture conditions to express their totipotency, acquire embryogenic characteristics, and eventually give rise to somatic embryos. When SE is initiated from centaury root explants, the process occurs spontaneously (on hormone-free medium), directly (without the callusing phase), and the somatic embryos are of unicellular origin. SE from leaf explants has to be induced by plant growth regulators and is indirect (preceded by callusing). Histological observations and culture conditions are compared in these two systems. The changes in antioxidative enzymes were followed during SE from the leaf explants. Special focus is given to the role of arabinogalactan proteins during SE, which were analyzed using a variety of approaches. The newest and preliminary results, including centaury transcriptome, novel potential SE markers, and novel types of arabinogalactan proteins, are discussed as perspectives of centaury research.

Keywords: antioxidative enzymes; arabinogalactan proteins; centaury; Gentianaceae; in vitro culture; morphogenesis; plant growth regulators; somatic embryo; tissue culture

Citation: Simonović, A.D.; Trifunović-Momčilov, M.M.; Filipović, B.K.; Marković, M.P.; Bogdanović, M.D.; Subotić, A.R. Somatic Embryogenesis in *Centaurium erythraea* Rafn—Current Status and Perspectives: A Review. *Plants* **2021**, *10*, 70.
https://doi.org/10.3390/plants10010070

Received: 15 December 2020
Accepted: 29 December 2020
Published: 31 December 2020

Publisher's Note: MDPI stays neutral with regard to jurisdictional claims in published maps and institutional affiliations.

1. Somatic Embryogenesis: Biotechnological Exploitation of Plant Cells' Totipotency

Plants have unique developmental plasticity, which allows their adaptation to constant environmental changes. Plant in vitro culture techniques relies on this plasticity to mold the morphogenetic paths in the desired direction. Morphogenetic processes enabling the regeneration of the whole plant in in vitro tissue culture conditions are somatic embryogenesis (SE), organogenesis, micropropagation, androgenesis, and gynogenesis. Differentiated somatic cells grown in vitro begin to divide and can regenerate the whole plant through SE or organogenesis [1]. SE is the process during which somatic cells, under inductive conditions, form embryogenic cells that undergo morphological and biochemical changes leading to the formation of a somatic embryo [2]. SE is a powerful biotechnological method for the propagation and genetic improvement of many plant species, as it enables the obtaining of a large number of somatic embryos, which can be further used in the production of artificial seeds with diverse applications in biotechnology [3].

Somatic embryos can develop from a wide range of differentiated cell types, such as ovule, embryo, root, leaf, and meristem cells, in response to different exogenous and/or endogenous factors [4]. The presence and level of endogenous factors (phytohormones) determine whether SE can occur spontaneously, on a hormone-free medium, or must be induced by the addition of plant growth regulators (PGRs). SE can be direct (DSE), without an intermediate callusing phase, or indirect (ISE), which implies the formation of disorganized callus tissue [5]. Somatic embryos developed through DSE or ISE may be of uni- or multicellular origin [6].

Somatic cells are not totipotent per se, and they need induction under appropriate conditions [7]. During the induction phase, somatic cells acquire embryogenic competence and proliferate, while during the expression phase, embryogenic cells differentiate into somatic embryos [8]. These two phases are thought to be mutually independent and influenced by different factors. Competent cells represent a transition from somatic to embryogenic state, which still requires exogenous stimuli, while the embryogenic cells have the ability to regenerate embryos without exogenous stimuli [8,9]. Inductive conditions, such as exogenously added PGRs and stress factors, lead to the dedifferentiation of plant cells and activation of the embryogenic pathway [10]. It is still unclear why and how differentiated plant cells become totipotent and acquire embryogenic potential and why this phenomenon occurs only in certain plant species, certain tissue types, or cells [10]. Many genes are involved in the vegetative to embryogenic transition. These include phytohormone-responsive genes, such as auxin-related and ABA-inducible genes, genes involved in the cell cycle control, genes involved in growth and remodeling of the cell wall, as well as an array of transcription factors. The involvement of *LEC* (leafy cotyledon), *BBM* (baby boom), *WUS* (wuschel), *CLV* (clavata), *STM* (shoot meristemless), *SHR* (short root), *ABI3* (abscisic acid-insensitive), *FUS3* (fusca) and other transcription factorshave been confirmed in SE regulation [11]. Somatic Embryogenesis Receptor-Like Kinases (SERKs) are well-known SE-specific signaling components [12]. Although significant progress in identifying factors involved in induction, perception, and signal transduction during SE has been made, the results of these numerous studies are still fragmentary and insufficient to explain the events occurring during SE at the molecular level.

While the initiation of embryogenic tissues depends on the developmental stage of the used initial plant material and components of the nutrient medium, the sustention of the embryogenic potential during subsequent cultivation requires the simultaneous activity of signaling and genetic pathways. Identification of proteins and genes involved in the control of the embryogenic potential of the plant cells represents one of the most efficient ways for understanding the molecular mechanisms of SE. These molecules, so-called SE markers, can allow the identification of the cells with embryogenic potential in the tissue culture before visible morphological changes. Several proteins isolated during SE are stress-related or pathogen-related proteins. These proteins were isolated from different plants during stress treatments of plant tissue culture imposed by wounding, desiccation, heavy metals, or PGRs [13]. Extracellular proteins are also potentially good markers of SE because they play an important role in plant cell differentiation [14]. The largest number of extracellular proteins are glycoproteins, of which arabinogalactan proteins (AGPs) are especially important during SE. The current review covers several aspects of SE, including the influence of explant type and culture conditions (PGRs and light conditions), as well as the roles of antioxidative enzymes and AGPs, investigated in a medicinal plant *Centaurium erythraea*.

2. Centuries of Centaury Research

Centaurium erythraea Rafn, commonly known as centaury, is a pharmacologically important medicinal plant from the Gentianaceae family. Centaury is a biennial, sometimes annual herb, which grows in wet to semi-arid areas throughout the northern hemisphere It is an ancient medicinal plant with the longest tradition in many pharmacopeias: it was described by Dioscorides nearly 2000 years ago. Centaury is used for the treatment of a wide range of ailments [15–17]. Using pure phytochemicals or crude plant extracts, experimental trials have been performed to evaluate anticancer, antioxidant, anti-inflammatory, antipyretic, analgesic, antimicrobial, antidiabetic, gastro-,cardio- and hepatoprotective activities of this important plant in many experimental animal systems [18–24]. A wide range of bioactive compounds can be found in the aerial part of *C. erythraea*, including secoiridoids, indole alkaloids, phenolic compounds (xanthones, flavonoids, and phenolic acid), and terpenes [25–27]. Centaur can also be used in the food processing industry as a natural flavoring or additive [26].

The ever-expanding demand for centaury in traditional medicine and the pharmaceutical industry has led to its uncontrolled collection resulting in a rapid decline of its natural populations. Development of methods for in vitro mass propagation of the centaury plants, as well as strategies for biotechnological production of its active metabolites, have attracted the attention of several research groups, resulting in a number of publications.

C. erythraea is relatively easily manipulated in the in vitro culture, where it can even complete its life cycle. Centaury's manageability and developmental plasticity in vitro made it not only the most investigated species of the *Centaurium* genus but an emerging model in plant developmental biology. The diversity of morphogenetic paths that *C. erythraea* can undergo in vitro has been compiled recently [28]. The vigorous morphogenic potential of the explants favors the use of centaury for genetic transformations [29–31], interspecific hybridization in vitro [32–34], functional studies on secondary metabolite synthesis in vitro [25,26,35–37], and stress physiology studies [38,39]. Overall, different aspects of *C. erythraea's* in vitro development, physiology, pharmacology, and ecology have been studied for over20 years at the Department for Plant Physiology, Institute for Biological Research "Siniša Stanković", University of Belgrade, resulting in 38 journal papers, 8 book chapters, and 9 masters and doctoral theses. Similarities and differences between two in vitro systems for the induction of SE in centaury that have been extensively studied in our lab—spontaneous DSE from root culture and induced ISE from leaf explants—are the focus of this review. A newly developed system for secondary and cyclic SE is also described [40] and will be submitted as an accompanying article in this issue.

3. SE from Centaury Root Explants Is Spontaneous and Direct

Both SE and organogenesis in vitro can proceed either spontaneously, on hormone-free media, or be induced by (a combination of) PGRs. PGRs exogenously added to the nutrition medium, as well as the content of endogenous hormones in different plant tissues, affect the induction of SE [8,9,41]. Auxins and cytokinins (CKs) are the main factors that determine the response and direction of SE by controlling the cell cycle, activation of cell divisions, and cell differentiation [5,8], so their presence is required for SE induction [42–44]. However, only certain auxins, such as 2,4-dichlorophenoxyacetic acid (2,4-D) or naphthaleneacetic acid (NAA), were key factors for inducing embryogenic cells from immature leaves of *Manihot esculenta* [45]. In addition, 2,4-D or picloram (PIC) induced direct SE in leaf segments of *Petiveria alliacea* [46]. On the other hand, there are reports describing that CKs, such as thidiazuron (TDZ) and 6-benzylaminopurine (BA), played a crucial role in SE form leaf and shoot explants of *Ochna integerrima* and leaf explants of *Primulina tabacum* [47,48].

The first successful induction of SE in centaury was obtained in a cell suspension derived from callus cultures [49]. The calli were initiated from roots and shoots of seedlings on medium supplemented with kinetin (KIN, 10^{-6} M) and auxins indole-3-acetic acid (IAA, 10^{-5} M) or 2,4-D (10^{-6} M). In contrast to 2,4-D, IAA showed a stimulatory effect on SE induction from the cell suspension, but light was the main embryogenesis-inducing factor in this system.

Subotić et al. [35] achieved the induction of SE from centaury root culture on solid half-strength Murashige and Skoog medium ($\frac{1}{2}$MS) without the addition of PGRs in the light. The somatic embryos developed alongside adventitious buds, so somatic embryos and adventitious buds of different developmental stages could be observed on the same root explant. Histological studies revealed that the somatic embryos formed directly from the epidermal cells, without the callusing phase, while adventitious buds developed from root cortex tissue [50]. In other words, SE from centaury root explants was spontaneous, direct (DSE), asynchronous, and occurred simultaneously with organogenesis. Somatic embryos derived from root explants of in vitro-grown centaury followed a unicellular pathway of DSE. These observations were in accordance with an earlier report describing in vitro morphogenesis from apical segments of primary hairy roots [29]. Even though the DSE from roots occurred spontaneously, the effects of PGRs added to the culture were further investigated. Subotić et al. [51] reported the effects of exogenous gibberellic acid (GA$_3$) and paclobutrazol, an inhibitor of gibberellin synthesis, both added at concentrations 0.01–3.0 µM, on the SE induction in wild type and hairy root centaury culture. It was shown that GA$_3$ had an inhibitory effect on the process of SE, while paclobutrazol in all applied concentrations had a stimulatory effect. The induction of SE in solid centaury root culture is presented in Figure 1.

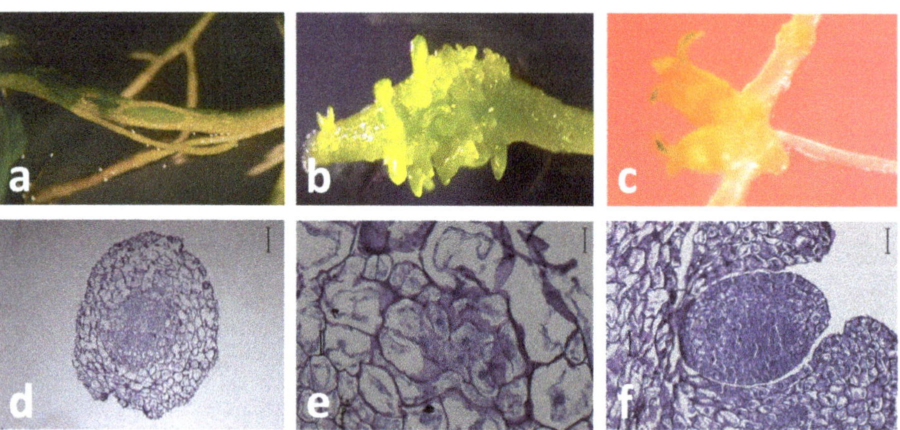

Figure 1. Direct somatic embryogenesis (SE) in *Centaurium erythraea* solid root culture. (**a**) The first response of root explant is enlargement and clear morphological changes observed five days after the culture setup on half-strength Murashige and Skoog ($\frac{1}{2}$MS) medium, (**b**) Detail of root explant with somatic embryos, (**c**) Cotyledonary somatic embryos developed directly from the root explant with no intervening callus phase, (**d**) Cross-section of root explant at the beginning of the culture. Scale bar indicates 200 µm, (**e**) Histological appearance of a proembryogenic structure. Scale bar indicates 100 µm, (**f**) Somatic embryo originated directly from the epidermal and subepidermal cells of the root tissue. Scale bar indicates 100 µm.

As recently reviewed by Tomiczak et al. [52], SE has also been successfully obtained from the root cultures of several other species from the Gentianacea family, even though in these cases, the SE was not spontaneous as in centaury but required the addition of PGRs. Mikuła and Rybczyńsky [53] tried to induce SE in *G. cruciate* root explants cultured on MS medium supplemented with 2,4-D and kinetin. The root explants formed callus tissue at the cut surface precisely on the wounding site of roots. Further ultrastructural analysis has shown that the structures originating from single cortical cells resembled proembryos in root explants of *G. cruciate*, but the process of SE was not further continued [54]. On the other hand, in *G. kurroo*, *G. pannonica*, and *G. cruciate*, somatic embryos were regenerated by rhizodermal cells of adventitious roots [43]. This process was stimulated by various combinations of auxins and CKs, and somatic embryos were further converted into plantlets on a $\frac{1}{2}$MS medium. Successful induction of SE was also achieved on root explants of *G. lutea* grown on a medium supplemented with auxins alone or in combination with cytokinin, although the conversion of somatic embryos into plantlets required the addition of mannitol or sorbitol to the basal culture medium [55]. The initiation of somatic embryos was also obtained in root explants of *Eustoma grandiflorum* cultured on a medium with 2,4-D [56]. The somatic embryos originated from pericycle and vascular parenchyma cells of seedling roots. Further conversion of the somatic embryos into plantlets was enabled with the addition of BA or GA$_3$ [56].

4. Indirect SE from Centaury Leaf Explants

The successful induction of SE, as well as shoot and root regeneration in vitro, depends on a variety of factors, including the explant selection, light conditions, and exogenously added PGRs [5,57]. The leaf culture, implying in vitro cultivation of isolated leaves, is generally not used because the whole leaves cannot be maintained in tissue culture. However, if the leaf sections are used as initial explants, then calli, buds, or somatic embryos can be induced since some mesophyll cells have the potential to re-enter the cell cycle and become committed to different morphogenetic pathways when appropriately induced. Leaves from in vitro cultivated plants are an easily accessible source of explants, while the leaf culture enables the regeneration of genetically stable plants [58].

The effect of nutrient media and different PGRs on regeneration possibilities from centaury leaf explants have been investigated in several previous studies [59–62], but in all these reports, only adventitious buds and calli regenerated on the leaf explants. Recent research revealed that centaury leaf explants cultured on hormone-free medium in the light produced only a few shoots, while roots developed in darkness [28]. Both organogenesis and rhizogenesis occurred directly, without the callusing stage, but no somatic embryos developed on hormone-free media [28].

The first successful induction of SE from the centaury leaf explants was reported by Filipović et al. [28] on media containing *N*-(2-chloro-4-pyridyl)-*N*′-phenylurea (CPPU) and 2,4-D, applied together, where the embryogenic response increased with the increasing CPPU concentration. Synthetic urea-type cytokinin CPPU has a diverse morphogenic activity in different species [63]; other tested PGRs with cytokinin activity—6-benzyladenine, kinetin, and thidiazuron—induced callus proliferation only [28]. This combination of PGRs(CPPU + 2,4-D) induced somatic embryo formation in *Gentiana* spp. leaf explants, as well [43]. When the centaury leaf explants are cultivated on CPPU and 2,4-D, the direction of morphogenesis depends on the light conditions: If the explants are cultivated in darkness, the indirect formation of somatic embryos (ISE) is the only process that occurs, but when the explants are kept in the light, the processes of ISE and indirect shoot development (ISD) proceeded simultaneously, and both were asynchronous [28]. Even though ISE can be isolated from other morphogenetic paths by culturing the explants in darkness, a higher frequency of embryogenic callus induction was obtained in the light. Thus, it can be concluded that in centaury leaf culture, light is an obligatory factor for the organogenesis, but also a factor that enhances ISE [28], which is in accordance with previous reports where light-induced SE in centaury suspension culture [49], as well as the frequency of SE and the number of embryos per leaf explant of *Dendrobium* [64] and *Petiveria alliacea* cultures [46]. The developed somatic embryos originated from the leaf subepidermal cells [28].

Plant regeneration via SE in leaf culture was also obtained in other gentian species. ISE in *G. pneumonanthe* was achieved on $\frac{1}{2}$MS supplemented with 2,4-D and BA [65]. The embryogenic potential of leaf explants was also investigated in *G. kurroo*, *G. cruciata*, *G. tibetica*, *G. lutea*, *G. pannonica* [43]. The leaf explants of these species were grown on a medium supplemented with three auxins and five different CKs, and optimum regeneration was achieved in the presence of NAA in combination with BA or TDZ (thidiazuron). Furthermore, cytomorphological analyses have shown that somatic embryos originated from palisade mesophyll cells. SE was also induced on leaf explants of *G. straminea* and *G. utriculosa* cultured on an MS medium supplemented only with 2,4-D [66,67]. On the other hand, in leaf explants of *G. straminea*, *G. macrophylla*, and *S. chirata*, successful induction of embryogenic callus was achieved on medium with a combination of 2,4-D and CKs [68–70]. The process of ISE from the centaury leaf explants is illustrated in Figure 2 and the Supplementary video.

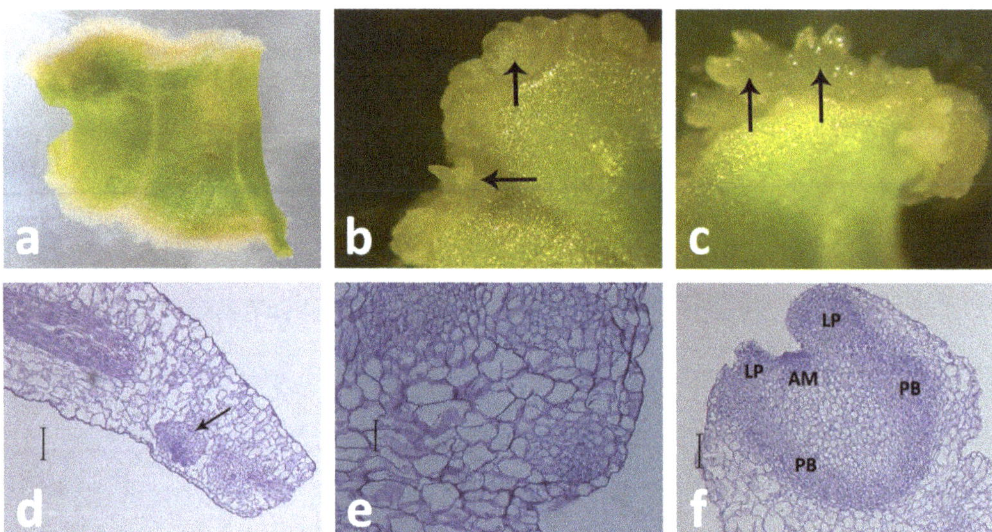

Figure 2. SE in *Centaurium erythraea* leaf culture. (**a**) Embryogenic callus developed at the edge of the leaf explant treated with 2,4-dichlorophenoxyacetic ac(2,4-D) and *N*-(2-chloro-4-pyridyl)-*N′*-phenylurea (CPPU) in darkness, (**b**) and (**c**) Somatic embryos at different stages of development (arrows), (**d–f**) Micrographs showing somatic embryo development on a leaf explant, (**d**) Histological appearance of a meristematic center (arrow) in the subepidermal layer of the leaf explant. Scale bar indicates 200 μm, (**e**) Globular somatic embryo. Scale bar indicates 100 μm, (**f**) Cotyledonary somatic embryo with apical meristem (AM), leaf primordial (LP), and provascular bundles (PB). Scale bar indicates 200 μm.

Since the two main systems for the SE induction in centaury, DSE from root culture [35] and ISE from leaf culture [28], differ in their requirements regarding the addition of PGRs for the SE induction, the endogenous contents of different CKs, IAA, salicylic acid (SA) and abscisic acid (ABA) were analyzed in the roots and shoots of the in vitro grown plants as the sources of explants [71]. It was found that the total amount of endogenous CKs was 1.4 times higher in shoots as compared to shoots, but inactive or weakly active *N*-glucosides were the predominate CK forms in both organs, whereas free bases and O-glucosides represented only a small portion of the total CK pool. The roots were characterized with higher IAA content but lower IAA/free CK bases ratio and lower ABA content in comparison to roots. The most significant difference, however, was a 44-fold higher SA content in the roots as compared to shoots [71]. It is not clear which of these differences allows spontaneous SE from roots but not from shoots; for example, Quiroz-Figueroa et al. [2] demonstrated that very low concentrations of salicylates could induce cellular growth and enhance somatic embryogenesis in *Coffea arabica.* Planned investigation considering the determination of endogenous levels of phytohormones at different stages of somatic embryo development aims to relate these levels to the embryogenic capacity of centaury root and shoot explants. The processes of SE from centaury root and leaf cultures are presented in Figure 3.

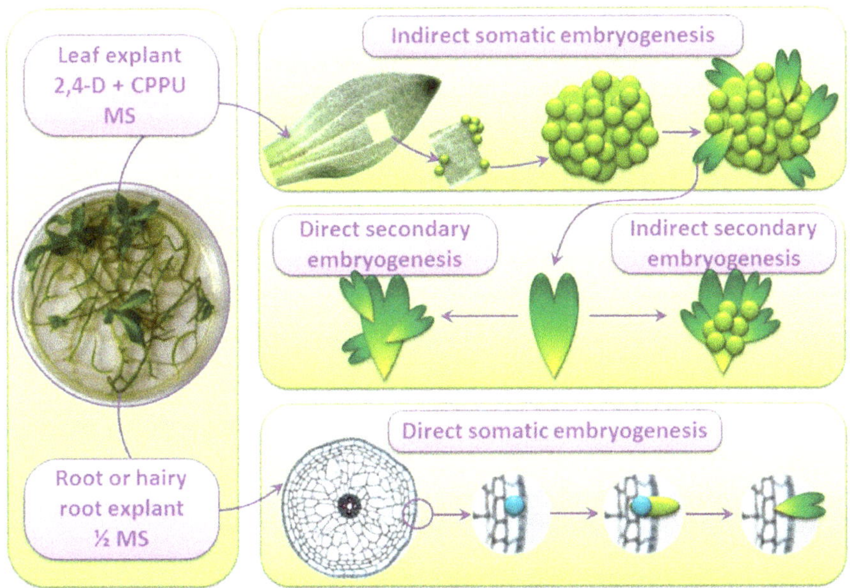

Figure 3. Schematic overview of SE in *Centaurium erythraea*. Left panel: Both leaves and roots of the in vitro grown *C. erythraea* plants can serve as sources of explants for the induction of SE. Hairy root cultures can also be used as explants. Upper panel: SE can be induced from leaf explants on the inductive medium containing 2,4-D and CPPU, both in the light and in darkness. Somatic embryos form from differentiated somatic cells in the subepidermal layer of the leaf explant. In this case, SE is indirect and proceeds via the callusing phase. Middle panel: The obtained somatic embryos can be further used as explants for secondary or cyclic embryogenesis [40] Lower panel: SE from root or hairy root explants is spontaneous, on $\frac{1}{2}$MS medium and direct. SE starts with asymmetric divisions of single totipotent cells from the epidermal or subepidermal layers of root explant. Successive divisions give rise to somatic embryos.

5. Maintaining Reactive Oxygen Species Homeostasis during SE in Centaury: The Role of Antioxidative Enzymes

Three decades ago, Dudits et al. [72] suggested that the somatic embryo initiation in vitro was a stress response. Many reports since then underlined the importance of stress factors during SE induction in vitro [10,73–76]. Cultured plant tissues experience a variety of stresses as a consequence of in vitro manipulations, including wounding, sterilization, mineral nutrient imbalance in the culture medium composition, PGRs, or subcultures. In response to any of these stresses, the homeostasis between reactive oxygen species (ROS) production and scavenging is disturbed, and ROS are generated in excess, thereby imposing oxidative stress in plant tissue culture. Stresses experienced by cultured tissues may induce a general response, resulting in chromatin remodeling and activation of the embryogenic developmental program [72,74]. Namely, accumulating evidence revealed that ROS (specifically H_2O_2) may function as signaling molecules that regulate plant growth and development, including cellular proliferation and differentiation [77,78]. As a cellular messenger, H_2O_2 in proper concentrations has the ability to induce gene expression and protein synthesis, hence triggering activation of embryogenic developmental program and formation of somatic embryos in different plant species [76]. On the other hand, excessive ROS could severely damage cellular proteins, DNA, and membrane lipids [78]. Thus, ROS overproduction could lead to plant recalcitrance and reduced morphogenetic competence during the in vitro culture [79]. Therefore, maintaining an optimum ROS level in the cell and restoring cell redox balance is important and enables the regulation of various processes [78], including SE induction.

The level of H_2O_2 is controlled by the activities of several key enzymes, including superoxide dismutases (SODs), catalases (CATs), and class III peroxidases (POXs) [80]. The SODs provide the front-line defense against ROS since they scavenge superoxide radicals to produce H_2O_2 [81]. CATs remove the excess of H_2O_2, while extracellular POXs play a role in the precise regulation of ROS levels in the cell and apoplast because, in addition to their role in removing H_2O_2, they can also catalyze the formation of H_2O_2 and hydroxyl radicals [82–84]. By regulating ROS levels in the apoplast, POXs participate in cross-linking, cell wall reconstruction, and elongation [81]. In many plants, these antioxidant enzymes have been shown to play an important role in scavenging ROS that arise during SE [75,77].

To our best knowledge, the only study on the roles of antioxidative enzymes in relation to SE within the Gentianaceae family is the study on the already described system of regeneration and ISE induction from centaury leaf explants [28]. Filipović et al. [28] investigated the activities of SODs, CATs, and POXs in a comprehensive set of samples comprising intact leaves, wounded explants, and explants grown either in light or darkness on three types of media, of which one inductive medium (0.2 mg/L 2,4-D and 0.5 mg/L CPPU) supported ISE. Of these, only the changes in the antioxidative activities in response to wounding and during ISE will be discussed here.

Wounding of the centaury leaves (cutting the leaves into explants) caused an increase in SOD activity (comprising 3 Cu/Zn-SOD isoforms), an increase in CAT activity (comprising 3 major activity bands), as well as a decrease in total POX activity [28], indicating that SOD and CAT are involved in the protection of centaury leaves from wounding-induced oxidative damage. Wounding leads to an accumulation of ROS in *Medicago truncatula* leaf explants, which occurs within seconds [85]. Slesak et al. [86] showed that mechanical injury of *Mesembryanthemum cristallinum* leaves leads to H_2O_2 accumulation, which was accompanied by an increase in total SOD activity and a decrease in CAT activity. Decreased POX activity in response to wounding, recorded in centaury leaves, is consistent with low POX activity in freshly isolated leaf explants of *Dactylis glomerata* [1]. Mechanical wounding is an inevitable consequence of in vitro manipulations. Wound signaling triggers not only defense responses, such as the production of ROS but also healing responses, including dedifferentiation, cell cycle reactivation, and vascular regeneration [87].

Following rapid responses to wounding, ROS homeostasis has to be reestablished, which is crucial for initial cell dedifferentiation and division during callus formation [88]. Subsequent planting of the centaury leaf explants on inductive medium strongly induced total POX activity, both in light and darkness, suggesting the importance of these enzymes in cell division, growth, and differentiation, probably through their action on cell wall remodeling [82]. A statistically significant increase in POX activity in comparison to the control intact leaves occurred after seven days of incubation, when the first cell divisions and the formation of meristem centers in the sections of the centaury leaves were observed, with the peaks of POX activity on the 14th or 21st day in culture, which coincidence with the emergence of somatic embryos. Therefore, it could be concluded that POXs play an important role in the development of centaury somatic embryos. Previous reports confirmed the important role of POX during SE induction from leaf explants of *D. glomerata* [1] and *Cicer arietinum* [44]. On the contrary, SOD activity decreased in light and remained unchanged during ISE in darkness, while CAT activity decreased during ISE both in light and darkness. The obtained results illustrate that dynamic changes in the antioxidative enzymatic capacity upon wounding and in response to SE induction are required to maintain ROS homeostasis in centaury leaf explants.

6. Studies on the Role of AGPs during SE in Centaury Using β-D-glucosyl Yariv Reagent

AGPs are heavily glycosylated, intrinsically disordered glycoproteins ubiquitous in plants, which belong to a superfamily of cell surface hydroxyproline-rich glycoproteins (HRGPs) [89,90]. The extraordinary structural diversity of AGPs relies not only on their protein backbones encoded by large gene families [89,91] but also on the possibility of differential glycosylation of the same isoform into hetero generous glycoforms [92]. Structural features that are common to AGPs include the presence of branched type II arabino-3,6-galactans (AGs) and short oligoarabinosides(both *O*-linked to the hydroxyproline (Hyp) residues), a high percentage of amino acids that constitute the AG-II glycomodules (Pro/Hyp, Ala, Ser, Thr, and Gly), N-terminal signal peptide directing their synthesis via secretory pathway, and often a C-terminal glycosylphosphatidylinositol (GPI) lipid anchor signal peptide [93,94]. While many AGPs are GPI-anchored to the plasma membranes, others may be secreted to the medium [93,95]. Dragićević et al. [89] recently demonstrated that many AGP sequences may have transmembrane domains. Beside these basic structural features' characteristic for classical AGPs and their short counterparts, AG peptides, many AGPs contain additional conserved domains or functional motifs and are termed chimeric AGPs [94].

AGPs are involved in cell proliferation [96] and diverse developmental and physiological processes, including differentiation and patterning [93,95,97]. Involvement of AGPs in SE has been described in many plant species, such as maize [98], chicory [99], *Trifolium nigrescens* [100], and others. As discussed below, the role of AGPs during SE from centaury roots [101,102] and leaf explants [103,104] has been investigated using a variety of approaches. One of the main tools for studying the AGPs' functions, used for decades, is a synthetic red dye, β-D-glucosyl Yariv reagent or βGlcY [105]. Most AGPs specifically bind βGlcY [1,3,5-tris (4-β-D-glycopyranosyloxyphenylazo)-2,4,6-trihydroxybenzene]; for β-galactosyl Yariv reagent, similar to βGlcY, a noncovalent interaction with β-1,3-galactan moieties of AGPs has been demonstrated [106].

βGlcY has been widely used as a histochemical reagent to detect AGPs int issue sections [107]. When *C. erythraea* roots are used for SE induction on a hormone-free medium, initially, the whole root explants were stained with βGlcY, but the most intense staining was in the epidermal cells and vascular tissue [101,102]. After one week in culture, βGlcY intensively stained AGPs in the surface cell layers of the centaury root explants, where somatic embryos were likely to develop. A similar staining pattern was observed in the outer epidermal cells during SE induction in chicory root culture [99]. Considering that somatic embryos originate directly from the root epidermal cells [50], the accumulation of AGPs in this region is indicative of their involvement in SE initiation. After two weeks in culture, the subepidermal layers of root explants also reacted with βGlcY, but neither developing globular embryos nor adventitious buds (which form alongside the embryos) showed significant precipitation of AGPs with βGlcY [102]. Finally, after 8 weeks in root culture, the epidermal and subepidermal cells were deeply stained with βGlcY, while staining of vascular tissue was less intense. This is shown in the root cross-section (Figure 4a), where several developed somatic embryos, as well as adventitious buds, can be seen.

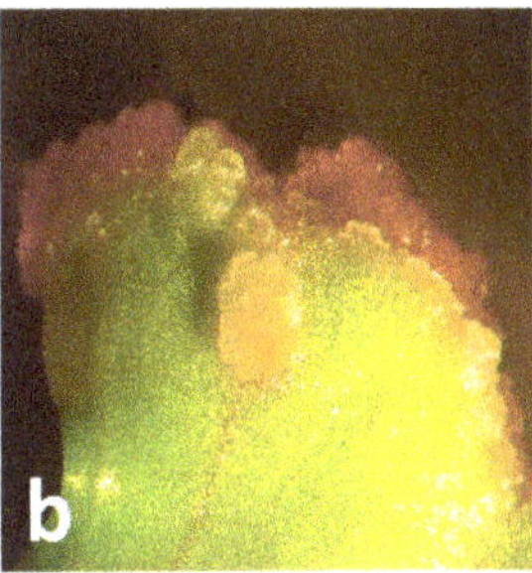
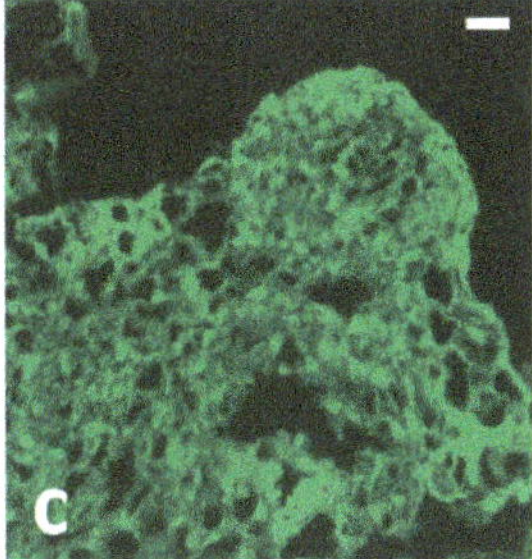

Figure 4. Distribution of AGPs during SE in centaury. (**a**) Cross-section of a root explant with somatic embryos at its surface, stained with βGlcY reagent. Scale bar indicates 80 μm (**b**) Indirect SE on centaury leaf explants grown on 100 μM βGlcY reagent in darkness. Somatic embryos form only on the parts of the explants that are not in direct contact with the medium. (**c**) Embryogenic globule developed on leaf explant and labeled with JIM15 antibody. Scale bar indicates 10 μm.

To study the role of AGPs during morphogenesis in vitro, βGlcY can be applied as an adjuvant to the culture medium [96,99,108,109]. Inactivation of AGPs by βGlcY binding during the induction of SE in different systems, commonly inhibits SE [99] and/or affects embryos' development and morphology [108,109]. As expected, the addition of βGlcY to the inductive medium during the induction of ISE from centaury leaf explants in darkness reduced the number of developed somatic embryos per explant in a dose-dependent manner [103]. The concentration of 150 μM βGlcY almost completely inhibited ISE, whereas, at lower concentrations, the embryos developed only in the explants' regions that were not in direct contact with the βGlcY-containing medium (Figure 4b). Indirect shoot development on the same inductive medium in the light or direct shoot development on a hormone-free medium were also inhibited by βGlcY in a concentration-dependent manner but were less sensitive to βGlcY than ISE [103]. The obtained results clearly point at AGPs as essential factors during ISE from centaury leaf explants.

However, quite unexpected results were obtained when βGlcY was used to investigate the role of AGPs during morphogenesis (simultaneous development of somatic embryos and adventitious buds) from centaury root explants [101,102]. Namely, it turned out that βGlcY may actually stimulate the morphogenesis from the root explants, albeit not in a linear dose-response manner. βGlcY increased the shoot regeneration frequency of roots cultured on the hormone-free medium from 71.67% for untreated culture to 93.89% and 92.22% for cultures grown on 15 μM or 25 μM βGlcY, respectively. The same concentrations also increased the average number of regenerated shoots per root explant, while lower (5 μM) or higher (50–75 μM) βGlcY concentrations had little effect on the regeneration potential of the 8-week-old root culture in comparison to untreated control [101]. Comparable results were obtained when the regeneration was scored after four weeks in culture [102] or when 1 μM IBA was added to the medium [100]. In any case, the obtained regenerants displayed normal morphology. Interestingly, the shoots regenerated on media containing 25–75 μM βGlcY had elevated content of AGPs in comparison to control shoots [101], as determined by the single radial gel diffusion method, which also utilizes βGlcY [110]. The roots developed on regenerated shoots also had increased AGP levels when developed on βGlcY-containing media [101]. This finding suggests that blocking of AGPs may increase their synthesis by some type of feedback regulation. The authors suggested that βGlcY in tissue culture may act as a stressor that may stimulate regeneration [102] since βGlcY triggers wound-like responses in *Arabidopsis* cell culture, as shown by whole-genome array [111]. On the other hand, the presence of 75 μM βGlcY in the centaury 4-weeks old leaf culture did not alter the AGP content, regardless of other conditions (basal or inductive medium and light vs. darkness) [103], so the effect of βGlcY on AGPs accumulation might be tissue-specific. Finally, the profile of AGPs present in the regenerating leaf explants, as determined by crossed electrophoresis [112], depends on the medium composition, light conditions, and culture age [103].

7. Dynamic Changes of AGPs Distribution and Expression during SE in Centaury

Overall, the evidence collected using βGlcY reagent in different assays suggests that AGPs are important for the induction of SE in centaury. However, tracking the dynamicsof AGPs' distribution during regeneration requires more sophisticated methods, such asa widely use dimmunohistochemical approach with monoclonal antibodies (mAbs) raised against AGPs' carbohydrate epitopes.A large set of the anti-AGP mAbs of the JIM, LM, and MAC series are commercially available and are listed, along with the epitopes they recognize, in several reviews [95,96]. The exact structure of the recognized epitopes is not always clearly determined, but it is known that MAC207, JIM4, and JIM13 bind to the β-D-GlcA-(1→3)-α-D-GalA-(1→2)-α-L-Rha motif, whereasLM2 recognizes β-linked glucuronic acid (β-D-GlcA). A systematic review of the available literature describing the expression of different mAb-recognizable AGP epitopes during SE in different species [98–100,109], to mention a few, would surely transcend the scope of the current review and would only confirm that none of the tested epitopes stands out as a universal SE marker in all or most of the studied plant species. Thus, the existence of two systems for SE induction in the same species—DSE on the hormone-free medium from root explants [102] and ISE from the leaf explants cultured on inductive medium [104]—provides an opportunity for the comparison of the obtained immunohistochemical results and search for common epitope markers or patterns.

The morphogenesis from the root explants was studied using LM2, JIM13, JIM15, JIM16, and MAC207 mAbs, of which the expression of JIM13-reactive epitope was not detected at all [102]. The JIM16 epitopes were localized in all cells of the root explants, especially in the endodermis and the central cylinder, as well as in the newly formed meristematic centers, so they were considered as markers of organogenesis, not somatic embryogenesis. The remaining three mAbs reacted with the epitopes present in somatic embryos. LM2 epitopes were widely distributed in root cross-sections at the beginning of the culture, but after 4 weeks in culture, the LM2 signal was more localized in epidermal cells and newly formed globular somatic embryos [102]. Comparable results for LM2 localization were obtained during DSE from chicory roots [99], where this epitope was foundin the surface cell layer surrounding somatic embryos; however, in this system, JIM13 and JIM16 mAbs were also expressed. MAC207 epitope had strong expression in protodermal cells of the embryos, as well as at the surface of epidermal cells of root explants adjacent to globular somatic embryos, with a strong signal in the extracellular matrix. Finally, the JIM15 epitope was reactive with AGPs in developed somatic embryos, as well as in the cells of the vascular elements of the root explants [102].

A slightly different set of mAbs, comprisingJIM4, JIM8, JIM13, JIM15, LM2, LM14, and MAC207, was used to investigate the distribution of the corresponding AGP epitopes during ISE from *C. erythraea* leaf explants [104]. As discussed above, in this system, light induces simultaneous development of both somatic embryos and adventitious buds, whereas in darkness, only ISE occurs. Generally, in globular somatic embryos, a different distribution pattern of JIM4, JIM13, JIM15, LM2, and MAC207 epitopes was observed, while with the progression of SE, the number of detected AGPs decreased. When the explants are cultivated in darkness, the JIM4 epitope was strongly expressed from the earliest stages of SE: It localized in the epidermal and subepidermal cells which formed meristematic centers, in the embryogenic cells in meristematic calli, as well as in four-cell proembryo. During further proembryo development, strong expression was detected only in the extracellular matrix surrounding the proembryogenic nodule. At the globular stage, JIM4 epitopes were found in the cell walls of the protodermal cells, while at the early cotyledonary stage, the JIM4 fluorescence was moderate [104]. Even though this mAb was not exclusively present in embryogenic tissues, its expression in adventitious buds formed in the light was weak. In maize callus culture, JIM4 was as also an early marker of embryogenic competence [98]. A similar labeling pattern to JIM4 in globular somatic embryos was found for MAC207 since its strong signal was detected in the cell walls of protodermal cells [104], just as was seen in protodermal cells of the globular embryos developed from roots [102].The JIM13 epitope, which was not observed during SE from the root explants at all [102], showed an intense signal in the whole globular embryos developed from leaf explants in darkness but was not restricted only to the embryogenic tissues [104]. The expression of JIM13 decreased in late embryos. In adventitious buds developed in the light, this epitope was not detected. Strong JIM13 labeling was also found in the embryogenic sector during SE in peach palm, where it was associated with extracellular matrix surface network [108]. Likewise, high-intensity JIM15 (Figure 4c) and LM2 fluorescence were localized to the whole globular embryos, but not in later developmental stages. A strong LM2 signal was observed in the cell walls of meristematic cells from which somatic embryos develop and in cells of embryogenic swellings in *Trifolium nigrescens* [100]. Both JIM15 and LM2 signals were also seen in developing adventitious buds, so they were not an exclusive feature of SE. Unlike other tested epitopes, which appeared early during ISE, the LM14 signal was not present in globular somatic embryos but was strong and evenly distributed throughout the longitudinal sections of the heart embryo. Finally, theJIM8 epitope was detected in the extracellular matrix, as well as in adventitious buds, but not in somatic embryos of any stage [104].

The obtained immunohistochemical results only corroborated well-established observations that spatiotemporal occurrence of AGPs during SE is developmentally regulated and that AGPs may serve as positional markers, markers of cell identity, or markers for embryogenic competence [97–100]. Our results indicate that the profile of AGP epitopes expressed during SE is not only species-specific but also strongly depends on the explant type and the culture conditions: While some epitopes, such as MAC207, have similar expression patterns in both regeneration systems, others, such asJIM13, are strongly expressed in somatic embryos developed from leaves, but are absent in embryos regenerated on *C. erythraea* roots. Furthermore, even though some mAbs recognize the same epitope (MAC207, JIM4, and JIM13), they display different labeling patterns.

Even though anti-AGP mAbs have been widely used for studying AGPs' distribution during SE and other developmental processes, their usefulness is intrinsically limited for several reasons: (1) mAbs are not specific for a single AGP; (2) they cannot distinguish all glycoforms of an AGP backbone, and (3) the epitope has to be unmasked for immunodetection [89,91,92]. Of course, that the analysis of the spatiotemporal pattern of gene expression can indicate the involvement of a particular AGP in some process [93], but in non-model species, such as *C. erythraea*, the necessary sequence resources are commonly unavailable. Thus, we initially used in-house assembled centaury leaf and root transcriptomes [113] and mined centaury AGP sequences using a homology-based search. Using this approach, we have identified four centaury AGP transcripts, named *CeAGP1* through *CeAGP4* [103]. Of these, *CeAGP1*, *CeAGP2*, and *CeAGP4* (GenBank: KC733882, KC733883, and KC733885, respectively) were characterized with conserved fasciclin domains and represented members of a subclass of chimeric AGPs known as fasciclin-like AGPs or FLAs [114]. *CeAGP1* was highly induced (26.7-fold) during morphogenesis from centaury leaf explants in the light, where ISE was accompanied with indirect shoot development, but more importantly, it was over 20 fold induced during ISE in darkness, in comparison to the control explants, indicating its importance during ISE [103]. *CeAGP2* was slightly induced during both direct (on hormone-free medium) and indirect morphogenetic paths (ISE and indirect shoot development on inductive media), while the induction of *CeAGP4* during ISE and organogenesis on inductive media was very low [103]. The role of *CeAGP1* in ISE can be viewed in light of the general role of FLAs as molecules involved in cell adhesion and protein–protein interactions [114]. *CeAGP3* is an AG peptide with a conserved DUF1070 domain (GenBank: KC733884, protein:AGN92423). The expression pattern of *CeAGP3* indicated its general involvement indifferent morphogenetic paths in centaury, since this transcript was induced 36.6-fold relative to control during ISE in darkness, but was also highly induced during indirect morphogenesis in the light (ISE and organogenesis), as well as in direct organogenesis on a hormone-free medium [103]. We have analyzed all 271 sequences containing the DUF1070 domain (DUF stands for Domain of Unknown Function) from 25 diverse families of vascular plants, aiming to elucidate the function of this motif. As it turned out, most of the DUF1070 domain represented typical glycosylphosphatidylinositol lipid anchor signal peptide (GPIsp) found in short AGPs (AG peptides), so the DUF1070 was renamed to arabinogalactan peptide (PF06376) [115]. To our best knowledge, DUF1070/PF06376 is the only conserved domain exclusively found in AGPs and HRGPs, in general. GPI anchors in proteins, such as *CeAGP3,* are proposed to increase lateral mobility of the anchored proteins in the plasma membrane, allow polarized targeting to the cell surface, inclusion in lipid rafts, as well as further processing by GPI-specific phospholipases and glycosidases, thereby releasing diffusible AGPs and/or carbohydrates as extracellular signals, as well as biologically active lipids as intracellular signals [91,94,95,97,103,115]; any of the proposed features for GPI-anchored AGPs may be important for morphogenesis and SE.

8. Perspectives: Novel SE Markers, "AGP-Tyr Kinases", and Time-Laps Embryogenesis

To support the analysis of the molecular events during SE and other in vitro morphogenetic processes in centaury, we have recently sequenced six *C. erythraea* transcriptomes (embryogenic calli, globular somatic embryos, cotyledonary somatic embryos, adventitious buds, leaves and roots of in vitro grown plants) and de novo assembled referent transcriptome comprising 105.726 genes [116]. The high quality and completeness transcriptome were functionally annotated and made publicly available. The transcriptome, along with a set of validated housekeeping genes, comprises a framework for the search for genes involved in SE and organogenesis [116]. A subset of genes potentially involved is SE was selected as transcripts with ≥8-fold higher expression (FPKM values) in embryogenic tissues as compared to non-embryogenic tissues, and their expression was further analyzed by qRT-PCR in 16 tissue samples [117]. The most intriguing finding of this preliminary research was the expression profile an unknown gene (provisionally termed *UN1*), a 725 bp long transcript with no BLAST hits or homology with any known sequence. *UN1* was highly expressed in leaf-derived embryogenic calli, while its expression progressively decreased in globular and cotyledonary embryos. The *UN1* expression in seedlings, roots, leaf-derived adventitious buds and leaves from flowering plants was below the qPCR detection limits, implying that its expression is restricted to the initial ISE stages [11]. The investigation of *UN1* structure and function is ongoing; for now, we can only speculate that *UN1* may have an impact on the acquisition of the embryogenic potential, and as such it may be a novel SE marker.

As discussed above, a homology-based search revealed only four AGP sequences in the first version of the *C. erythraea* transcriptome, all of which were AGPs with conserved domains [103]. This was expected since HRGPs, including AGPs, are intrinsically disordered proteins lacking hydrophobic core, so the sequence constraints imposed on these proteins are relatively low. Therefore, AGPs can rapidly mutate and evolve, which hinders their homology-based mining [90]. We have recently developed a highly sophisticated bioinformatics pipeline developed in R, ragp, for mining and analysis of HRGPs with an emphasis on AGPs [90]. The key novelty incorporated in ragp is the machine learning-based prediction of proline hydroxylation sites, which represent the glycosylation sites. The analysis of *C. erythraea* transcriptome [118] as well as 62 plant proteomes using ragp [90] revealed, quite unexpectedly, that the most frequently identified domains found in AGPs were the Protein kinase and Protein tyrosine kinase domains. The Protein (tyrosine) kinase domains have thus far eluded experimental evidence for linkage with AGPs in any plant species. Possible implications of this finding include a novel way of attachment of AGPs to the plasm membrane through their transmembrane domains and a novel way for the involvement of AGPs in signaling. So far, structural features of AGPs and circumstantial evidence suggested that AGPs may be involved in signaling as co-receptors [97], or through interaction with membrane receptors (including protein kinases) on the same or neighboring cell [11,91,93], interaction with other AGPs, pectins, and other cell wall or cytoskeletal elements [89,93]. The presence of protein kinase domains on ragp-predicted AGPs suggests that AGPs may actually be the membrane receptors themselves or that certain Protein kinases have previously undetected AG-glycomodules and can be glycosylated. While the experimental evidence for the Hyp-glycosylation of these protein kinases is lacking, and the functions of the proposed "AGP-Protein kinase" molecules are unknown, it should be noted that, for example, *A. thaliana* SERK5 (AT2G13800.1) has predicted hydroxyprolines organized in characteristic AG-glycomodules [90,118] (Figure 5). The analysis of expression and function of "AGP-Protein kinases" and their possible involvement in SE in *C. erythraea* is planned.

Since both DSE from centaury roots [35,50] and ISE from leaf explants [28] are asynchronous, collecting embryogenic tissues, specifically somatic embryos at different developmental stages for molecular and biochemical analyses, is very tedious and time-consuming. Unfortunately, the establishment of a synchronized embryogenic culture has not been achieved in *C. erythraea* yet, and it remains one of our goals. A synchronized culture would not only aid the harvest of somatic embryos at a specific stage but would also indicate the exact timing to a specific developmental event under certain conditions. An alternative way to achieve this is documentation of the development of embryogenic structures on selected explants over time. Such documentation system has been established by the combination of photography (using a smartphone camera with a macro lens), image processing of focalstacks from the developing explants automated in Adobe Photoshop and Bridge, and a relational database built using Excel and R [119]. An example of such a time-lapse documentation video of SE from leaf explants is provided as a Supplement.

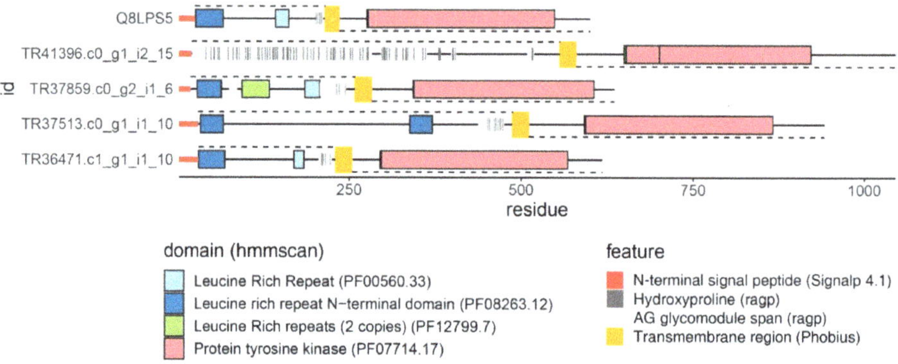

Figure 5. Arabinogalactan proteins (AGPs) with Tyr kinase domains. The first sequence is Somatic Embryogenesis Receptor-like Kinase 5 (SERK5) from *A. thaliana* (AT2G13800.1 or Q8LPS5 protein precursor). Four sequences below are found in the *C. erythraea* transcriptome, based on homology with SERK5. In addition to AG glycomodules with predicted hydroxyprolines and Tyr kinase domains, all sequences have N-terminal signal peptide and a transmembrane domain, while most have Leucine-rich repeats typical for SERK receptors. The image is generated using ragp.

9. Conclusions

Even after 20 years of research, *Centaurium erythraea* remains an attractive, challenging, and yet rewarding experimental object at our Department. *C. erythraea* is already firmly established as a valued model system at our Department for the studies on alternative ways for the production of secondary metabolites and for the studies on morphogenesis in vitro—both primarily aimed at its conservation and sustainable usage. However, the accumulated data and successful protocols for the centaury propagation in vitro [28] have led us to gradually shift our focus from centaury's potential as a medicinal plant to its possibly even greater potential as a genetic resource for crop improvement. Namely, we believe that centaury's immense regeneration potential and developmental plasticity, when cultivated in vitro, rely on the presence or high activity of certain genes that may not be present or active in plant species recalcitrant to SE induction or in vitro propagation and manipulations in general. Having a sequenced *C. erythraea* transcriptome [116] would allow us, and other research groups interested in centaury development, to mine for genes that are highly active during SE and organogenesis, and hopefully find genes, such as *UN1* [117], that were not described before. Such novel genes, as well as known genes previously unassociated with morphogenesis, could be considered as sequence resources for the genetic improvement of valuable crops that are recalcitrant to in vitro manipulations. In addition, finding genes that are differentially expressed during DSE and ISE from roots and leaves, respectively, specifically AGP genes and genes associated with auxin and cytokinin signaling or metabolism, as well as differences in the endogenous hormones during these two processes, would probably highlight some factors governing SE via direct or indirect pathway. Thus, unrevealing at least a part of the molecular networks and genes that are at the base of SE induction and other regeneration processes in centaury is the primary focus for our future research.

Supplementary Materials: The following are available online at https://www.mdpi.com/2223-774 7/10/1/70/s1, Time laps photo-documentation of somatic embryogenesis from *Centaurium erythraea* leaf culture is provided in a form of a video file Centaury somatic embryogenesis.mp4. The size of the video is about 25 Mb.

Author Contributions: Conceptualization, A.R.S., A.D.S., M.M.T.-M., and B.K.F.; writing—original draft preparation, Sections 1 and 2, A.R.S.; writing—original draft preparation, Sections 3 and 4, M.M.T.-M. and M.P.M.; writing—original draft preparation, Section 5, B.K.F.; writing—original draft preparation, Sections 6 and 7, A.D.S.; writing—original draft preparation, Section 8, A.D.S. and M.D.B.; writing—review and editing, A.D.S.; visualization, all authors.; Supplementary video, M.D.B.; supervision, A.R.S. All authors have read and agreed to the published version of the manuscript.

Funding: This research received no external funding.

Acknowledgments: This work was supported by the Ministry of Education, Science and Technological Development of the Republic of Serbia, contract number 451-03-68/2020-14/200007. The authors are grateful to Milan Dragićevićfor Figure 5 generated in ragp. The authors also thank Milica Simonović for rendering and post-production of the somatic embryogenesis time-laps video (Supplement).

Conflicts of Interest: The authors declare no conflict of interest.

References

1. Somleva, M.; Alexieva, V.; Sergiev, I.; Karanov, E. Alterations in the activities of some hydrogen peroxide scavenging enzymes during induction of somatic embryogenesis in leaf explants from *Dactylis glomerata* L. *Dokl. Bolg. Akad. Nauk.* **2000**, *53*, 91–94.
2. Quiroz-Figueroa, F.; Méndez-Zeel, M.; Larqué-Saavedra, A.; Loyola-Vargas, V. Picomolar concentrations of salicylates induce cellular growth and enhance somatic embryogenesis in *Coffea arabica* tissue culture. *Plant Cell Rep.* **2001**, *20*, 679–684. [CrossRef]
3. Rihan, H.Z.; Kareem, F.; El-Mahrouk, M.E.; Fuller, M.P. Artificial seeds (principle, aspects and applications). *Agronomy* **2017**, *7*, 71. [CrossRef]
4. Hortsman, A.; Bemer, M.; Boutilier, K. A transcriptional view on somatic embryogenesis. *Regeneration* **2017**, *4*, 201–216. [CrossRef]
5. Phillips, G.C. In vitro morphogenesis in plants-recent advances. *In Vitro Cell. Dev. Biol. Plant* **2004**, *40*, 342–345. [CrossRef]
6. Quiroz-Figueroa, F.R.; Rafael, R.H.; Galaz-Avalos, R.M.; Loyola-Vargas, V.M. Embryo production through somatic embryogenesis can be used to study cell differentiation in plants. *Plant Cell Tissue Organ Cult.* **2006**, *86*, 258–301. [CrossRef]
7. Fehér, A. Callus, dedifferentiation, totipotency, somatic embryogenesis: What these terms mean in the era of molecular plant biology? *Front Plant Sci.* **2019**, *10*, 536. [CrossRef]
8. Jimenez, V.M. Involvement of plant hormones and plant growth regulators on in vitro somatic embryogenesis. *Plant Growth Regul.* **2005**, *47*, 91–110. [CrossRef]
9. Fehér, A.; Pasternak, T.P.; Dudits, D. Transition of somatic plant cells to an embryogenic state. *Plant Cell Tissue Organ Cult.* **2003**, *74*, 201–228. [CrossRef]
10. Fehér, A. Somatic embryogenesis—Stress-induced remodeling of plant cell fate. *Biochim. Biophys. Acta* **2015**, *1849*, 385–402. [CrossRef]
11. Smertenko, A.; Bozhkov, P.V. Somatic embryogenesis: Life and death processes during apical-basal patterning. *J. Exp. Bot.* **2014**, *65*, 1343–1360. [CrossRef]
12. Karami, O.; Saidi, A. The molecular basis for stress-induced acquisition of somatic embryogenesis. *Mol. Biol. Rep.* **2010**, *37*, 2493–2507. [CrossRef] [PubMed]
13. Gulzar, B.; Mujib, A.; Moien Qadir, M.; Sayeed, R.; Mamgain, J.; Ejaz, B. Genes, proteins and other networks regulating somatic embryogenesis in plants. *J. Genet. Eng. Biotechnol.* **2020**, *18*, 31. [CrossRef] [PubMed]
14. Tchorbadjieva, M.I. Protein markers for somatic embryogenesis. In *Plant Cell Monographs*; Mujib, A., Šamaj, J., Eds.; Springer: Berlin/Heidelberg, Germany, 2005; pp. 215–233. [CrossRef]
15. Haloui, M.; Louedec, L.; Michel, J.B.; Lyoussi, B. Experimental diuretic effects of *Rosmarinus officinalis* and *Centaurium erythraea*. *J. Ethnopharmacol.* **2000**, *71*, 465–472. [CrossRef]
16. Stefkov, G.; Miova, B.; Dinevska-Kjovkarovska, S.; Stanoeva, J.P.; Stefova, M.; Petrusevska, G.; Kulevanova, S. Chemical characterization of *Centaurium erythrea* L. and its effects on carbohydrate and lipid metabolism in experimental diabetes. *J. Ethnopharmacol.* **2014**, *152*, 71–77. [CrossRef] [PubMed]
17. Guedes, L.; Reis, P.B.P.S.; Machuqueiro, M.; Ressaissi, A.; Pacheco, R.; Serralheiro, M.L. Bioactivities of *Centaurium erythraea* (Gentianaceae) decoctions: Antioxidant activity, enzyme inhibition and docking studies. *Molecules* **2019**, *24*, 3795. [CrossRef] [PubMed]

18. Berkan, T.; Ustünes, L.; Lermioglu, F.; Ozer, A. Antiinflammatory, analgesic, and antipyretic effects of an aqueous extract of *Erythraea centaurium*. *Planta Med.* **1991**, *57*, 34–37. [CrossRef]

19. Valentão, P.; Fernandes, E.; Carvalho, F.; Andrade, P.B.; Seabra, R.M.; Bastos, M.L. Antioxidant activity of Centaurium erythraeainfusion evidenced by its superoxide radical scavenging and xanthine oxidase inhibitory activity. *J. Agric. Food Chem.* **2001**, *49*, 3476–3479. [CrossRef]

20. Kumarasamy, Y.; Naha, L.; Cox, P.J.; Jaspars, M.; Sarker, S.D. Bioactivity of secoiridoid glycosides from *Centaurium erythraea*. *Phytomedicine* **2003**, *10*, 344–347. [CrossRef]

21. Tuluce, Y.; Ozkol, H.; Koyuncu, I.; Ine, H. Gastroprotective effect of small centaury (*Centaurium erythraea* L.) on aspirin-induced gastric damage in rats. *Toxicol. Ind. Health* **2011**, *27*, 760–768. [CrossRef]

22. Trifunović-Momčilov, M.; Krstić-Milošević, D.; Trifunović, S.; Podolski-Renić, A.; Pešić, M.; Subotić, A. Secondary metabolite profile of transgenic centaury (*Centaurium erythraea* Rafn) plants, potential producers of anticancer compounds. In *Transgenesis and Secondary Metabolism, Reference Series in Phytochemistry*; Jha, S., Ed.; Springer International Publishing: Berlin/Heidelberg, Germany, 2016. [CrossRef]

23. Trifunović-Momčilov, M.; Krstić-Milošević, D.; Trifunović, S.; Ćirić, A.; Glamočlija, J.; Jevremović, S.; Subotić, A. Antimicrobial activity, antioxidant potential and total phenolic content of transgenic *AtCKX1* centaury (*Centaurium erythraea* Rafn) plants grown in vitro. *Environ. Eng. Manag. J.* **2019**, *18*, 2063–2072. [CrossRef]

24. Đorđević, M.; Grdović, N.; Mihailović, M.; Arambašić-Jovanović, J.; Uskoković, A.; Rajić, J.; Đordjević, M.; Tolić, A.; Mišić, D.; Šiler, B.; et al. *Centaurium erythraea* methanol extract protects red blood cells from oxidative damage in streptozotocin-induced diabetic rats. *J. Ethnopharmacol.* **2017**, *202*, 172–183. [CrossRef] [PubMed]

25. Šiler, B.; Avramov, S.; Banjanac, T.; Cvetković, J.; Nestorović-Živković, J.; Patenković, M.; Mišić, D. Secoiridoid glycosides as a marker system in chemical variability estimation and chemotype assignment of *Centaurium erythraea* Rafn from the Balkan Peninsula. *Ind. Crop. Prod.* **2012**, *40*, 336–344. [CrossRef]

26. Šiler, B.; Živković, S.; Banjanac, T.; Cvetković, J.; Nestorović-Živković, J.; Ćirić, A.; Soković, M.; Mišić, D. Centauries as underestimated food additives: Antioxidant and antimicrobial potential. *Food Chem.* **2014**, *147*, 367–376. [CrossRef] [PubMed]

27. Jovanović, O.; Radulović, N.; Stojanović, G.; Palić, R.; Zlatković, B.; Gudžić, B. Chemical composition of the essential oil of *Centaurium erythraea* Rafn (Gentianaceae) from Serbia. *J. Essent. Oil Res.* **2009**, *21*, 317–322. [CrossRef]

28. Filipović, B.K.; Simonović, A.D.; Trifunović, M.M.; Dmitrović, S.S.; Savić, J.M.; Jevremović, S.B.; Subotić, A.R. Plant regeneration in leaf culture of *Centaurium erythraea* Rafn Part 1: The role of antioxidant enzymes. *Plant Cell Tissue Organ Cult.* **2015**, *121*, 703–719. [CrossRef]

29. Subotić, A.; Budimir, S.; Grubišić, D.; Momčilović, I. Direct regeneration of shoots from hairy root cultures of *Centaurium erythraea* inoculated with *Agrobacterium rhizogenes*. *Biol. Plantarum* **2003**, *47*, 617–619. [CrossRef]

30. Trifunović, M.; Cingel, A.; Simonović, A.; Jevremović, S.; Petrić, M.; Dragićević, I.Č.; Motyka, V.; Dobrev, P.I.; Zahajská, L.; Subotić, A. Overexpression of *Arabidopsis* cytokinin oxidase/dehydrogenase genes *AtCKX1* and *AtCKX2* in transgenic *Centaurium erythraea* Rafn. *Plant Cell Tissue Organ Cult.* **2013**, *115*, 139–150. [CrossRef]

31. Trifunović, M.; Motyka, V.; Cingel, A.; Subotić, A.; Jevremović, S.; Petrić, M.; Holík, J.; Malbeck, J.; Dobrev, P.I.; Dragićević, I.Č. Changes in cytokinin content and altered cytokinin homeostasis in *AtCKX1* and *AtCKX2*-overexpressing centaury (*Centaurium erythraea* Rafn) plants grown in vitro. *Plant Cell Tissue Organ Cult.* **2015**, *120*, 767–777. [CrossRef]

32. Banjanac, T.; Šiler, B.; Skorić, M.; Ghalawenji, N.; Milutinović, M.; Božić, D.; Mišić, D. Interspecific in vitro hybridization in genus *Centaurium* and identification of hybrids via flow cytometry, RAPD, and secondary metabolite profiles. *Turk. J. Bot.* **2014**, *38*, 68–79. [CrossRef]

33. Banjanac, T.; Dragićević, M.; Šiler, B.; Gašić, U.; Bohanec, B.; Nestorović-Živković, J.; Trifunović, S.; Mišić, D. Chemodiversity of two closely related tetraploid *Centaurium* species and their hexaploid hybrid: Metabolomic search for high-resolution taxonomic classifiers. *Phytochemistry* **2017**, *140*, 27–44. [CrossRef]

34. Banjanac, T.; Đurović, S.; Jelić, M.; Dragićević, M.; Mišić, D.; Skorić, M.; Nestorović-Živković, J.; Šiler, B. Phenotypic and genetic variation of an interspecific *Centaurium* hybrid (*Gentianaceae*) and its parental species. *Plants* **2019**, *8*, 224. [CrossRef]

35. Subotić, A.; Janković, T.; Jevremović, S.; Grubišić, D. Plant Tissue Culture and Secondary Metabolites Productions of *Centaurium erythraea* Rafn, a Medical plant. In *Floriculture, Ornamental and Plant Biotechnology: Advances and Topical Issues*, 1st ed.; Teixeira da Silva, J.A., Ed.; Global Science Books: London, UK, 2006; pp. 564–570.

36. Subotić, A.; Jevremović, S.; Grubišić, D.; Janković, T. Spontaneous plant regeneration and production of secondary metabolites from hairy root cultures of *Centaurium erythraea* Rafn. In *Protocols for In Vitro Cultures and Secondary Metabolite Analysis of Aromatic and Medicinal Plants, Methods in Molecular Biology*; Jain, S.M., Saxena, P.K., Eds.; Springer: Berlin/Heidelberg, Germany, 2009; pp. 205–215. [CrossRef]

37. Božunović, J.; Živković, S.; Gašić, U.; Glamočlija, J.; Ćirić, A.; Matekalo, D.; Šiler, B.; Soković, M.; Tešić, Ž.; Mišić, D. In Vitro and In Vivo transformations of *Centaurium erythraea* secoiridoid glucosides alternate their antioxidant and antimicrobial capacity. *Ind. Crop. Prod.* **2018**, *111*, 705–721. [CrossRef]

38. Šiler, B.; Mišić, D.; Filipović, B.; Popović, Z.; Cvetić, T.; Mijović, A. Effects of salinity on in vitro growth and photosynthesis of common centaury (*Centaurium erythraea* Rafn). *Arch. Biol. Sci.* **2007**, *59*, 129–134. [CrossRef]

39. Trifunović-Momčilov, M.; Paunović, D.; Milošević, S.; Marković, M.; Jevremović, S.; Dragićević, I.Č.; Subotić, A. Salinity stress response of non-transformed and *AtCKX* transgenic centaury (*Centaurium erythraea* Rafn) shoots and roots grown in vitro. *Ann. Appl. Biol.* **2020**, *177*, 74–89. [CrossRef]

40. Bogdanović, M.; Ćuković, K.; Dragićević, M.; Simonović, A.; Subotić, A.; Todorović, S. Secondary somatic embryogenesis in *Centaurium erythraea*. In Proceedings of the 3rd International Conference on Plant biology, Belgrade, Serbia, 9–12 June 2018; p. 34.

41. Gutiérrez-Mora, A.; González-Gutiérrez, A.G.; Rodriguez-Garay, B.; Ascencio-Cabral, A.; Li-Wei, L. Plant Somatic Embryogenesis: Some useful considerations. In *Embryogenesis*; Ken-Ichi, S., Ed.; IntechOpen: London, UK, 2012. [CrossRef]

42. Fiuk, A.; Rybczyński, J.J. Morphogenic capability of *Gentiana kuroo* Royle seedling and leaf explants. *Acta Physiol. Plant.* **2008**, *30*, 157–166. [CrossRef]

43. Fiuk, A.; Rybczyński, J.J. Genotype and plant growth regulator dependent response of somatic embryogenesis from *Gentiana* spp. leaf explants. *In Vitro Cell. Dev. Biol. Plant* **2008**, *44*, 90–99. [CrossRef]

44. Ghanti, S.K.; Sujata, K.G.; Rao, S.; Udayakumar, M.; Kavi Kishor, P.B. Role of enzymes and identification of stage-specific proteins in developing somatic embryos of chickpea (*Cicer arietinum* L.). *In Vitro Cell. Dev. Biol. Plant* **2009**, *45*, 667–672. [CrossRef]

45. Ma, G.; Xu, Q. Induction of somatic embryogenesis and adventitious shoots from immature leaves of cassava. *Plant Cell Tissue Organ Cult.* **2002**, *70*, 281–288. [CrossRef]

46. Cantelmo, L.; Soares, B.O.; Rocha, L.P.; Pettinelli, J.A.; Callado, C.H.; Mansur, E.; Casteller, A.; Gagliardi, R.F. Repetitive somatic embryogenesis from leaves of the medicinal plant *Petiveria alliacea* L. *Plant Cell Tissue Organ Cult.* **2013**, *115*, 385–393. [CrossRef]

47. Ma, G.; Lu, J.; Teixeira da Silva, J.A.; Zhang, X.; Zhao, J. Shoot organogenesis and somatic embryogenesis from leaf and shoot explants of *Ochna integerrima* (Lour). *Plant Cell Tissue Organ Cult.* **2011**, *104*, 157–162. [CrossRef]

48. Yang, X.; Lu, J.; Teixeira da Silva, J.M.; Ma, G. Somatic embryogenesis and shoot organogenesis from leaf explants of *Primulina tabacum*. *Plant Cell Tissue Organ Cult.* **2012**, *109*, 213–221. [CrossRef]

49. Barešová, H.; Kamínek, M. Light induce embryogenesis in suspension culture of *Centaurium erythraea* Rafn. In *Plant Tissue and Cell Culture Propagation to Crop Improvement*; Novák, F.J., Havel, L., Doležel, J., Eds.; Czechoslovak Academy of Sciences: Prague, Czechoslovakia, 1984; pp. 163–164.

50. Subotić, A.; Grubišić, D. Histological analysis of somatic embryogenesis and adventitious formation from root explants of *Centaurium erythreae* Gillib. *Biol. Plantarum* **2007**, *51*, 514–516. [CrossRef]

51. Subotić, A.; Jevremović, S.; Trifunović, M.; Petrić, M.; Milošević, S.; Grubišić, D. The influence of gibberelic acid and paclobutrazol on induction of somatic embryogenesis in wild type and hairy root cultures of *Centaurium erythraea* Gillib. *Afr. J. Biotechnol.* **2009**, *8*, 3223–3228.

52. Tomiczak, K.; Mikuła, A.; Niedziela, A.; Wójcik-Lewandowska, A.; Domżalska, L.; Rybczyński, J.J. Somatic embryogenesis in the family Gentianaceae and its biotechnological application. *Front. Plant Sci.* **2019**, *10*, 762. [CrossRef]

53. Mikuła, A.; Rybczyńsky, J.J. Somatic embryogenesis of *Gentiana* genus I. The effect of the preculture treatment and primary explant origin on somatic embryogenesis of *Gentiana cruciata* (L.), *G. pannonica* (Scop.), and *G. tibetica* (King). *Acta Physiol. Plant* **2001**, *23*, 15–25. [CrossRef]

54. Mikuła, A.; Tykarska, T.; Rybczyńsk, J.J.; Kuraś, M. Ultrastructural analysis of initial stages of dedifferentiation of root explants of Gentianaseedlings. *Acta Soc. Bot. Pol.* **2002**, *71*, 287–297. [CrossRef]

55. Holobiuc, I. Somatic embryogenesis in long-term cultures of *Gentiana lutea* L. in the presence of osmotic stress. In *The Gentianaceae—Volume 2: Biotechnology and Application*; Rybczyński, J.J., Davey, M.R., Mikuła, A., Eds.; Springer: Berlin/Heidelberg, Germany, 2015; pp. 139–161. [CrossRef]

56. Yumbla-Orbes, M.; Ferreira da Cruz, A.C.; Marques Pinheiro, M.V.; Rocha, D.I.; Batista, D.S.; Koehler, A.D.; Barbosa, J.G.; Otoni, W.C. Somatic embryogenesis and de novo shoot organogenesis can be alternatively induced by reactivating pericycle cells in Lisianthus (*Eustoma grandiflorum* (Raf.) Shinners) root explants. *In Vitro Cell. Dev. Biol. Plant* **2017**, *53*, 209–218. [CrossRef]

57. Gaj, D.M. Factors influencing somatic embryogenesis induction and plant regeneration with particular reference to *Arabidopsis thaliana* (L.) Heynh. *Plant Growth Regul.* **2004**, *43*, 27–47. [CrossRef]

58. Thu, H.T.M.; Naing, A.H.; Jeong, H.Y.; Kim, C.K. Regeneration of genetically stable plants from in vitro vitrified leaves of different carnation cultivars. *Plants* **2020**, *9*, 950. [CrossRef]

59. Čellárová, E.; Repčáková, K.; Repčák, M.; Hončariv, R. Morphogenesis in tissue cultures of some medicinal plants. *Acta Hortic.* **1983**, *132*, 249–256. [CrossRef]

60. Barešová, H.; Herben, T.; Kamínek, M.; Krekule, J. Hormonal control of morphogenesis in leaf segments of *Centaurium erythraea*. *Biol. Plantarum* **1985**, *27*, 286–291. [CrossRef]

61. Laureová, D.; Čellárová, E.; Hončariv, R. Tollerance of plant tissue of *Centaurium erythraea* to increased concentrations of ions present in soils Eastern Slovakian lowlands. In *Dni Rastlinnej Fyziológie IV*; Repčák, M., Ed.; Slovenska Botaničká Spoločnost Pri Sav: Bratislava, Slovakia, 1986; pp. 221–222.

62. Piatczak, E.; Wysokinska, H. In vitro regeneration of *Centaurium erythraea* Rafn from shoot tips and other seedling explants. *Acta Soc. Bot. Pol.* **2003**, *72*, 283–288. [CrossRef]

63. Subotić, A.; Jevremović, S.; Grubišić, D. Influence of cytokinins on in vitro morphogenesis in root cultures of Centaurium erythraea—valuable medicinal plant. *Sci. Hortic.* **2009**, *120*, 386–390. [CrossRef]

64. Chung, H.H.; Chen, J.T.; Chang, W.C. Plant regeneration through direct somatic embryogenesis from leaf explants of *Dendrobium*. *Biol. Plantarum* **2007**, *51*, 346–350. [CrossRef]

65. Bach, A.; Pawłowska, B. Somatic embryogenesis in *Gentiana pneumonanthe* L. *Acta Biol. Cracov. Bot.* **2003**, *45*, 79–86.
66. Cai, Y.; Liu, Y.; Liu, Z.; Zhang, F.; Xiang, F.; Xia, G. High-frequency embryogenesis and regeneration of plants with high content of gentiopicroside from Chinese medicinal plant *Gentiana straminea* Maxim. *In Vitro Cell. Dev. Biol. Plant* **2009**, *45*, 730–739. [CrossRef]
67. Vinterhalter, B.; Mitić, N.; Vinterhalter, D.; Uzelac, B.; Krstić-Milošević, D. Somatic embryogenesis and in vitro shoot propagation of *Gentiana utriculosa*. *Biologia* **2016**, *71*, 139–148. [CrossRef]
68. He, T.; Yang, L.; Zhao, Z. Embryogenesis of *Gentiana straminea* and assessment of genetic stability of regenerated plants using inter simple sequence repeat (ISSR) marker. *Afr. J. Biotechnol.* **2011**, *10*, 7604–7610. [CrossRef]
69. Chen, L.Y.; Chen, Q.L.; Xu, D.; Hao, J.G.; Schläppi, M.; Xu, Z.Q. Changes of gentiopicroside synthesis during somatic embryogenesis in *Gentiana macrophylla*. *Planta Med.* **2009**, *75*, 1618–1624. [CrossRef]
70. Jha, B.T.; Dafadar, A.; Chaudhur, R.K. Somatic embryogenesis in *Swertiachirata* Buch. Ham. Ex Wall—A multipotent medicinal plant. *Asian J. Biotechnol.* **2011**, *3*, 186–193. [CrossRef]
71. Trifunović-Momčilov, M.; Motyka, V.; Dragićević, I.Č.; Petrić, M.; Jevremović, S.; Malbeck, J.; Holík, J.; Dobrev, P.I.; Subotić, A. Endogenous phytohormones in spontaneously regenerated *Centaurium erythraea* Rafn plants grown in vitro. *J. Plant Growth Regul.* **2016**, *35*, 543–552. [CrossRef]
72. Dudits, D.; Bogre, L.; Gyorgyey, J. Molecular and cellular approaches to the analysis of plant embryo development from somatic cells in vitro. *J. Cell Sci.* **1991**, *99*, 473–482.
73. Karami, O.; Aghavaisi, B.; Pour, A.M. Molecular aspects of somatic-to-embryogenic transition in plants. *J. Chem. Biol.* **2009**, *2*, 177–190. [CrossRef] [PubMed]
74. Zavattieri, M.A.; Frederico, A.M.; Lima, M.; Sabino, R.; Amhold-Schmidt, B. Induction of somatic embryogenesis as an example of stress-related plant reaction. *Electron. J. Biotechnol.* **2010**, *13*, 1–13. [CrossRef]
75. Tchorbadjieva, M.I. Advances in proteomics of somatic embryogenesis. In *Somatic Embryogenesis in Ornamentals and Its Applications*; Mujib, A., Ed.; Springer: Berlin/Heidelberg, Germany, 2016; pp. 67–90. [CrossRef]
76. Prudente, D.O.; de Souza, L.; Paiva, R. Plant somatic embryogenesis: Modulatory role of oxidative stress. *Proc. Natl. Acad. Sci. India Sect. B Biol. Sci.* **2019**, *90*, 483–487. [CrossRef]
77. Libik, M.; Konieczny, R.; Pater, B.; Slesak, I.; Miszalski, Z. Differences in the activities of some antioxidant enzymes and H_2O_2 content during rhizogenesis and somatic embryogenesis in callus cultures of the ice plant. *Plant Cell Rep.* **2005**, *2*, 834–841. [CrossRef]
78. Mittler, R. ROS are good. *Trends Plant Sci.* **2017**, *22*, 11–19. [CrossRef]
79. Benson, E.E. Special Symposium: In vitro plant recalcitrance do free radicals have a role in plant tissue culture recalcitrance? *In Vitro Cell. Dev. Biol. Plant* **2000**, *36*, 163–170. [CrossRef]
80. Apel, K.; Hirt, H. Reactive oxygen species: Metabolism, oxidative stress, and signal transduction. Reactive oxygen species:metabolism, oxidative stress, and signal transduction. *Annu. Rev. Plant Biol.* **2004**, *55*, 373–399. [CrossRef]
81. Alscher, R.G.; Erturk, N.; Heath, L.S. Role of superoxide dismutases (SODs) in controlling oxidative stress in plants. *J. Exp. Bot.* **2002**, *53*, 1331–1341. [CrossRef] [PubMed]
82. Liszkay, A.; Kenk, B.; Schopfer, P. Evidence for the involvement of cell wall peroxidase in the generation of hydroxyl radicals mediating extension growth. *Planta* **2003**, *217*, 658–667. [CrossRef] [PubMed]
83. Passardi, F.; Penel, C.; Dunand, C. Performing the paradoxical: How plant peroxidases modify the cell wall. *Trends Plant Sci.* **2004**, *9*, 534–540. [CrossRef] [PubMed]
84. Cosio, C.; Dunand, C. Specific functions of individual class III peroxidase genes. *J. Exp. Bot.* **2009**, *60*, 391–408. [CrossRef]
85. Wang, X.D.; Nolan, K.E.; Irwanto, R.R.; Sheahan, M.B.; Rose, R.J. Ontogeny of embryogenic callus in *Medicago truncatula*: The fate of the pluripotent and totipotent stem cells. *Ann. Bot.* **2011**, *107*, 599–609. [CrossRef]
86. Slesak, I.; Slesak, A.; Libik, M.; Libik, M. Antioxidant response system in the short-term post-wounding effect in *Mesembryanthemum crystallinum* leaves. *J. Plant Physiol.* **2008**, *165*, 127–137. [CrossRef]
87. Lup, S.D.; Tian, X.; Xu, J.; Pérez-Pérez, J.M. Wound signaling of regenerative cell reprogramming. *Plant Sci.* **2016**, *250*, 178–187. [CrossRef]
88. Rose, R.J. Somatic embryogenesis in the *Medicago truncatula* model: Cellular and molecular mechanisms. *Front. Plant. Sci.* **2019**, *10*, 267. [CrossRef]
89. Showalter, A.M.; Keppler, B.; Lichtenberg, J.; Gu, D.; Welch, L.R. A bioinformatics approach to the identification, classification, and analysis of hydroxyproline-rich glycoproteins. *Plant Phys.* **2010**, *153*, 485–513. [CrossRef]
90. Dragićević, M.B.; Paunović, D.M.; Bogdanović, M.D.; Todorović, S.I.; Simonović, A.D. ragp: Pipeline for mining of plant hydroxyproline-rich glycoproteins with implementation in R. *Glycobiology* **2020**, *30*, 19–35. [CrossRef]
91. Schultz, C.J.; Johnson, K.L.; Currie, G.; Bačić, A. The classical arabinogalactan protein gene family of Arabidopsis. *Plant Cell* **2000**, *12*, 1751–1767. [CrossRef] [PubMed]
92. Hijazi, M.; Durand, J.; Pichereaux, C.; Pont, F.; Jamet, E.; Albenne, C. Characterization of the Arabinogalactan Protein 31 (AGP31) of Arabidopsis thaliana new advances on the hyp-o-glycosylation of the pro-rich domain. *J. Biol. Chem.* **2012**, *287*, 9623–9632. [CrossRef] [PubMed]
93. Ellis, M.; Egelund, J.; Schultz, C.J.; Bačić, A. Arabinogalactan-proteins: Key regulators at the cell surface? *Plant Phys.* **2010**, *153*, 403–419. [CrossRef] [PubMed]

94. Nguema-Ona, E.; Coimbra, S.; Vicré-Gibouin, M.; Mollet, J.C.; Driouich, A. Arabinogalactan proteins in root and pollen-tube cells: Distribution and functional aspects. *Ann. Bot.* **2012**, *110*, 383–404. [CrossRef]

95. Showalter, A.M. Arabinogalactan-proteins: Structure, expression and function. *Cell. Mol. Life Sci.* **2001**, *58*, 1399–1417. [CrossRef]

96. Serpe, M.D.; Nothnagel, E.A. Effects of Yariv phenylglycosides on Rosa cell suspensions: Evidence for the involvement of arabinogalactan-proteins in cell proliferation. *Planta* **1994**, *193*, 542–550. [CrossRef]

97. Seifert, G.J.; Roberts, K. The biology of arabinogalactan proteins. *Annu. Rev. Plant Biol.* **2007**, *58*, 137–161. [CrossRef]

98. Šamaj, J.; Baluška, F.; Bobák, M.; Volkmann, D. Extracellular matrix surface network of embryogenic units of friable maize callus contains arabinogalactan-proteins recognized by monoclonal antibody JIM4. *Plant Cell Rep.* **1999**, *18*, 369–374. [CrossRef]

99. Chapman, A.; Blervacq, A.S.; Vasseur, J.; Hilbert, J.L. Arabinogalactan-proteins in Cichorium somatic embryogenesis: Effect of β-glucosyl Yariv reagent and epitope localisation during embryo development. *Planta* **2000**, *211*, 305–314. [CrossRef]

100. Pilarska, M.; Knox, J.P.; Konieczny, R. Arabinogalactan-protein and pectin epitopes in relation to an extracellular matrix surface network and somatic embryogenesis and callogenesis in *Trifolium nigrescens* Viv. *Plant Cell Tissue Organ Cult.* **2013**, *115*, 35–44. [CrossRef]

101. Trifunović, M.; Tadić, V.; Petrić, M.; Jontulović, D.; Jevremović, S.; Subotić, A. Quantification of arabinogalactan proteins during in vitro morphogenesis induced by β-D glucosyl Yariv reagent in *Centaurim erythraea* root culture. *Acta Physiol. Plant.* **2014**, *36*, 1187–1195. [CrossRef]

102. Trifunović, M.; Subotić, A.; Petrić, M.; Jevremović, S. The Role of Arabinogalactan Proteins in Morphogenesis of *Centaurium erythraea* Rafn In Vitro. In *The Gentianaceae—Volume 2: Biotechnology and Applications*; Rybczyński, J.J., Davey, M.R., Mikuła, A., Eds.; Springer: Berlin/Heidelberg, Germany, 2015; pp. 113–138. [CrossRef]

103. Simonović, A.D.; Filipović, B.K.; Trifunović, M.M.; Malkov, S.N.; Milinković, V.P.; Jevremović, S.B.; Subotić, A.R. Plant regeneration in leaf culture of *Centaurium erythraea* Rafn Part 2: The role of arabinogalactan proteins. *Plant Cell Tiss Organ Cult.* **2015**, *121*, 721–739. [CrossRef]

104. Filipović, B.K.; Trifunović-Momčilov, M.M.; Simonović, A.D.; Jevremović, S.B.; Milošević, S.M.; Subotić, A.R. Immunolocalization of some arabinogalactan protein epitopes during indirect somatic embryogenesis and shoot organogenesis in leaf culture of centaury (*Centaurium erythraea* Rafn). *In Vitro Cell. Dev. Biol. Plant.* **2021**, in press. [CrossRef]

105. Yariv, J.; Lis, H.; Katchalski, E. Precipitation of arabic acid and some seed polysaccharides by glycosylphenylazo dyes. *Biochem. J.* **1967**, *105*, 1C–2C. [CrossRef] [PubMed]

106. Kitazawa, K.; Tryfona, T.; Yoshimi, Y.; Hayashi, Y.; Kawauchi, S.; Antonov, L.; Tanaka, H.; Takahashi, T.; Kaneko, S.; Depree, P.; et al. β-Galactosyl Yariv reagent binds to the β-1, 3-galactan of arabinogalactan proteins. *Plant Phys.* **2013**, *161*, 1117–1126. [CrossRef]

107. Anderson, R.L.; Clarke, A.E.; Jermyn, M.A.; Knox, R.B.; Stone, B.A. A carbohydrate-binding arabinogalactan-protein from liquid suspension cultures of endosperm from *Lolium multiflorum*. *Aust. J. Plant Phys.* **1977**, *4*, 143–158. [CrossRef]

108. Thompson, H.J.; Knox, J.P. Stage-specific responses of embryogenic carrot cell suspension cultures to arabinogalactan protein-binding β-glucosyl Yariv reagent. *Planta* **1998**, *205*, 32–38. [CrossRef]

109. Steinmacher, D.A.; Saare-Surminski, K.; Lieberei, R. Arabinogalactan proteins and the extracellular matrix surface network during peach palm somatic embryogenesis. *Phys. Plant.* **2012**, *146*, 336–349. [CrossRef]

110. Van Holst, G.J.; Clarke, A.E. Quantification of arabinogalactan-protein in plant extracts by single radial gel diffusion. *Anal. Biochem.* **1985**, *148*, 446–450. [CrossRef]

111. Guan, Y.; Nothnagel, E.A. Binding of arabinogalactan proteins by Yariv phenylglycoside triggers wound-like responses in Arabidopsis cell cultures. *Plant. Phys.* **2004**, *135*, 1346–1366. [CrossRef]

112. Van Holst, G.J.; Clarke, A.E. Organ-specific arabinogalactan-proteins of Lycopersiconperuvianum (Mill) demonstrated by crossed electrophoresis. *Plant Physiol.* **1986**, *80*, 786–789. [CrossRef] [PubMed]

113. Malkov, S.; Simonović, A. Shotgun Assembly of *Centaurium erythraea* Transcriptome. In Proceedings of the 19th Symposium of the Serbian Plant Physiology Society, Banja Vrujci, Serbia, 13–15 June 2011; p. 16.

114. Johnson, K.L.; Jones, B.J.; Bačić, A.; Schultz, C.J. The fasciclin-like arabinogalactan proteins of Arabidopsis. A multigene family of putative cell adhesion molecules. *Plant Physiol.* **2003**, *133*, 1911–1925. [CrossRef] [PubMed]

115. Simonović, A.D.; Dragićević, M.B.; Bogdanović, M.D.; Trifunović-Momčilov, M.M.; Subotić, A.R.; Todorović, S.I. DUF1070 as a signature domain of a subclass of arabinogalactan peptides. *Arch. Biol. Sci.* **2016**, *68*, 737–746. [CrossRef]

116. Ćuković, K.; Dragićević, M.; Bogdanović, M.; Paunović, D.; Giurato, G.; Filipović, B.; Subotić, A.; Todorović, S.; Simonović, A. Plant regeneration in leaf culture of *Centaurium erythraea* Rafn Part 3: De novo transcriptome assembly and validation of housekeeping genes for studies of in vitro morphogenesis. *Plant Cell Tissue Organ Cult.* **2020**, 1–17. [CrossRef]

117. Ćuković, K.; Dragićević, M.; Todorović, S.; Bogdanović, M.B.; Simonović, A. Selection of differentially expressed genes in *Centaurium erythraea* Rafn during in vitro somatic embryogenesis. In Proceedings of the X International Scientific Agriculture Symposium AgroSym 2019, Jahorina, Republika Srpska, 3–6 October 2019; p. 226.

118. Paunović, D.; Bogdanović, M.; Trifunović-Momčilov, M.; Todorović, S.; Simonović, A.; Subotić, A.; Dragićević, M. Are receptor tyrosine kinases chimeric AGP's? In Proceedings of the 3rd International Conference on Plant Biology, Belgrade, Serbia, 9–12 June 2018; p. 17.

119. Bogdanović, M.; Ćuković, K.; Dragićević, M.; Simonović, A.; Todorović, S. Somatic embryogenesis of *Centaurium erythraea* Rafn time-lapse documentation of in vitro development. In Proceedings of the IX International Scientific Agriculture Symposium AgroSym 2018, Jahorina, Bosnia and Herzegovina, 4–7 October 2018; p. 347.

Technical Note

A ClearSee-Based Clearing Protocol for 3D Visualization of *Arabidopsis thaliana* Embryos

Ayame Imoto [1], Mizuki Yamada [2], Takumi Sakamoto [3], Airi Okuyama [3], Takashi Ishida [2], Shinichiro Sawa [4] and Mitsuhiro Aida [2,*]

1   Graduate School of Biological Sciences, Nara Institute of Science and Technology (NAIST), 8916-5 Takayama, Nara 630-0192, Japan; iabs.hr6@gmail.com
2   International Research Organization for Advanced Science and Technology (IROAST), Kumamoto University, 2-39-1 Kurokami, Chuo-ku, Kumamoto 860-8555, Japan; myamada@kumamoto-u.ac.jp (M.Y.); ishida-takashi@kumamoto-u.ac.jp (T.I.)
3   Faculty of Science, Kumamoto University, 2-39-1 Kurokami, Chuo-ku, Kumamoto 860-8555, Japan; t.s613@icloud.com (T.S.); flute.airi0709@gmail.com (A.O.)
4   Graduate School of Science and Technology, Kumamoto University, 2-39-1 Kurokami, Chuo-ku, Kumamoto 860-8555, Japan; sawa@kumamoto-u.ac.jp
*   Correspondence: m-aida@kumamoto-u.ac.jp; Tel.: +81-96-342-3402

Abstract: Tissue clearing methods combined with confocal microscopy have been widely used for studying developmental biology. In plants, ClearSee is a reliable clearing method that is applicable to a wide range of tissues and is suitable for gene expression analysis using fluorescent reporters, but its application to the *Arabidopsis thaliana* embryo, a model system to study morphogenesis and pattern formation, has not been described in the original literature. Here, we describe a ClearSee-based clearing protocol which is suitable for obtaining 3D images of *Arabidopsis thaliana* embryos. The method consists of embryo dissection, fixation, washing, clearing, and cell wall staining and enables high-quality 3D imaging of embryo morphology and expression of fluorescent reporters with the cellular resolution. Our protocol provides a reliable method that is applicable to the analysis of morphogenesis and gene expression patterns in *Arabidopsis thaliana* embryos.

Keywords: clearing; 3D imaging; *Arabidopsis thaliana*; embryo; confocal microscopy; cell wall staining; fluorescent reporter; GFP

Citation: Imoto, A.; Yamada, M.; Sakamoto, T.; Okuyama, A.; Ishida, T.; Sawa, S.; Aida, M. A ClearSee-Based Clearing Protocol for 3D Visualization of *Arabidopsis thaliana* Embryos. *Plants* **2021**, *10*, 190. https://doi.org/10.3390/plants10020190

Received: 31 December 2020
Accepted: 18 January 2021
Published: 20 January 2021

Publisher's Note: MDPI stays neutral with regard to jurisdictional claims in published maps and institutional affiliations.

1. Introduction

In plant development, oriented cell division and expansion play essential roles in morphogenesis and pattern formation [1]. The embryogenesis of *Arabidopsis thaliana*, in which a relatively small number of tissues and organs are arranged in a simple pattern, provides an excellent system to study morphogenesis and pattern formation, and many regulatory factors that affect these processes have been identified and studied extensively [2,3]. Moreover, because patterns of cell division and elongation are significantly regular during *Arabidopsis* embryogenesis [4–6], their possible roles in development and the underlying mechanisms for oriented cell division and elongation have been an important subject [7–10].

Because morphogenesis and pattern formation occur not only on the surfaces of the embryo but also in its internal structures (e.g., vascular and ground tissues), a reliable method for visualizing morphological and patterning events that occurs deep inside the embryo is necessary. Tissue clearing is a powerful technique to meet such requirements, and several protocols for clearing plant structures have been reported [11–14]. Among them, TOMEI-II [13] and ClearSee [11] have an advantage in visualizing gene expression patterns, as these methods well preserve the fluorescence of various fluorescent proteins. Although both methods can be applicable to a wide range of tissue types and to various plant species, whether they can also work with embryos has not been reported. Moreover, a recent application of ClearSee to *Arabidopsis thaliana* ovules [15] confirms that the method has potential to visualize plant internal structures with high quality. Here, we established a protocol to apply the ClearSee method to the embryo of *Arabidopsis thaliana* and to demonstrate that the protocol can visualize cellular arrangement and the signal of various fluorescent reporters in 3D.

2. Results and Discussion

2.1. Dissection of Embryos

We first applied the original ClearSee protocol [11] to whole seeds with the expectation of visualizing embryos without dissection. However, seeds processed with this protocol exhibited brown color in the endothelium (Figure 1A), preventing us from imaging internal embryos. We therefore decided to manually dissect embryos before applying the protocol.

For dissecting embryos, seeds were first removed from the fruit according to the method described previously [16] except with 7% glucose solution instead of N5T medium. Briefly, each of the valves was slit open using a needle and was partly removed from the fruit by using forceps to expose the seeds. The half-opened fruit was completely immersed in 3 mL of 7% glucose solution in a 35-mm dish, and seeds were excised by using a pair of forceps under a stereo microscope. To avoid floating of the seeds, which made the subsequent embryo isolation difficult, the whole fruit and hence the seeds within were kept submerged during dissection by holding its pedicel with another pair of forceps.

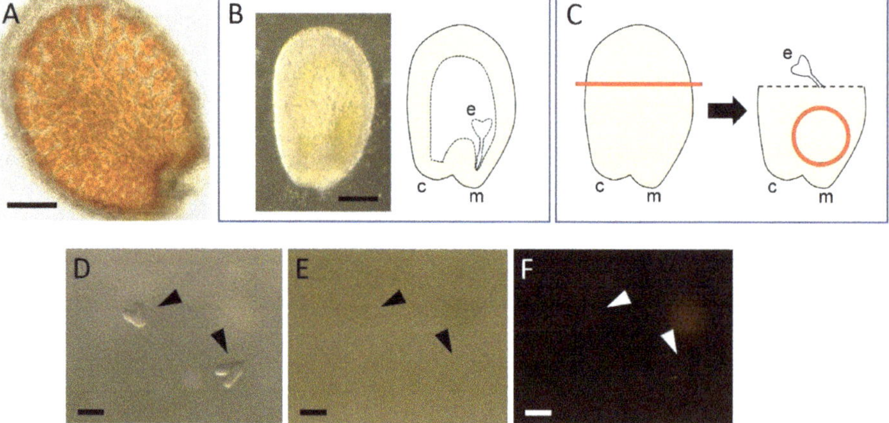

Figure 1. Dissection of *Arabidopsis thaliana* embryos: (**A**) a seed processed with the original ClearSee protocol [11], not transparent and exhibiting a brown color; (**B**) seeds excised from a fruit in 7% glucose solution (**left**) and schematic diagram of its internal structure (**right**); (**C**) the procedure of embryo isolation, where half of the seed is excised along the red line (**left**) and the region around the micropyle marked with a red circle (**right**) is pushed several times until the embryo pops out; and (**D–F**) the effects of illumination of a stereo microscope. In ClearSee solution, embryos are clearly visible with oblique transmitted illumination (**D**) whereas they are almost invisible with bright-field (**E**) or dark-field (**F**) illumination. The arrowheads indicate the positions of embryos. c, chalaza; e, embryo; and m, micropyle. Bars = 100 μm. Diagrams in B and C are modified from Hughes, 2009 [17].

The excised seeds were then subjected to manual dissection of embryos. Within a seed, the embryo is located on the micropyle/chalaza side (Figure 1B). To isolate embryos, the other side of the seed was excised by using forceps and the seed surface around the micropyle was gently pushed several times with the tips of the forceps until the embryo popped out from the open end (Figure 1C). Isolating 5–10 embryos each time, they were collected using a P20 micropipette, which was adjusted to 1–2 μL, and were assembled within a small area in the same 35-mm dish, where the debris produced by dissection was not present. This process was important to avoid losing isolated embryos by mixing them with the debris. Occasionally, small debris may adhere to an embryo, reducing the quality of imaging. Such debris can often be removed by gently scratching it with an eyelash attached to a toothpick.

2.2. Fixation and Washing

After tens of embryos were isolated and assembled, they were subjected to fixation and washing by transferring embryos from one solution to another by a P20 micropipette. To minimize the risk of losing embryos during this process, we put a drop of solution (100–200 μL) at the center of a dish instead of filling up the dish with a larger amount of solution. Moreover, embryos often adhered to the bottom of the dish or to the inner surface of the pipette tips when they were in the fixative or the washing buffer, further increasing the risk of losing them. To avoid this, we added a very small amount of ClearSee to each solution (e.g., 0.5 μL per 1 mL) prior to use because ClearSee contained the detergent sodium deoxycholate, which prevented embryos from adhering. When transferring embryos, the volume of the micropipette should be adjusted to 1–2 μL to minimize carryover of the solution.

The isolated embryos were collected from 7% glucose by using P20 micropipette under the stereo microscope and were transferred to a drop of the fixative. The embryos were incubated in the fixative for 10 min at room temperature. We tested 5, 10, 30, and 60 min as the fixation time, which gave essentially the same results in terms of embryo morphology, cell wall staining patterns, and fluorescence of reporters. The fixed embryos were then washed twice by sequentially transferring them to the first and second drops of the washing buffer that were placed in separate 35-mm dishes and by incubating them for 1 min each. Vacuum infiltration, which was described in the original ClearSee protocol [11], was not necessary.

2.3. Clearing and Staining

Clearing was carried out by transferring the fixed and washed embryos to 3 mL of ClearSee solution in a 35-mm dish. When the embryos were released from the P20 micropipette, they initially floated at the surface of the solution and then gradually sank until they reached the bottom as infiltration proceeded. The dish was then sealed with parafilm and was kept dark for 1–7 days at room temperature. Embryos at the late stages required longer incubation times compared to those at early stages.

From this step on, the embryos became difficult to see with the progression of clearing. To visualize the embryos for subsequent processing, the illumination settings of the stereo microscope were critical. Off-axis (oblique) illumination [18] gave significantly higher contrast than bright- or dark-field illumination, facilitating monitoring and collection of the embryo samples (Figure 1D–F).

The cleared embryos were then transferred to the staining solution containing Calcofluor White and were kept for 1 h at room temperature. Again, a 100–200 μL drop of the staining solution was used to avoid loss of embryo samples. After staining, the embryos were transferred to ClearSee and kept for 1 h to remove excess of the dye.

2.4. Confocal Microscopy

For mounting embryo samples, two pieces of double-sided tape were pasted with an appropriate interval onto a glass slide as spacers. The cleared embryos were mounted in an area between the spacers and covered with a coverslip. Marking the positions of the mounted embryos with a felt-tip pen on the coverslip helped to locate embryos under a confocal microscope.

Figure 2 shows a set of images obtained from an embryo carrying the *DR5rev::GFP* reporter [19]. Z-stack images of 157 serial optical sections with the 0.3-μm interval were acquired (Supplementary Movie S1) and used for observing a single focal plane (Figure 2A) or for reconstructing 3D image (Figure 2B,C). Both cell walls labelled by Calcofluor White and auxin response marked by the accumulation of endoplasmic reticulum

(ER)-localized Green Fluorescent Protein (GFP) are clearly visible, and both patterns of cellular configuration and distribution of the *DR5rev* activity are confirmatory with previous observations [8,20], showing that the cell wall structure and GFP fluorescence are well preserved after the processing with our protocol. Moreover, 3D reconstruction allows for identifying geometrical features of cell morphology and gene expression patterns. The image would also be suitable for quantitative analyses using imaging software such as ImageJ [21] or MorphographX [22].

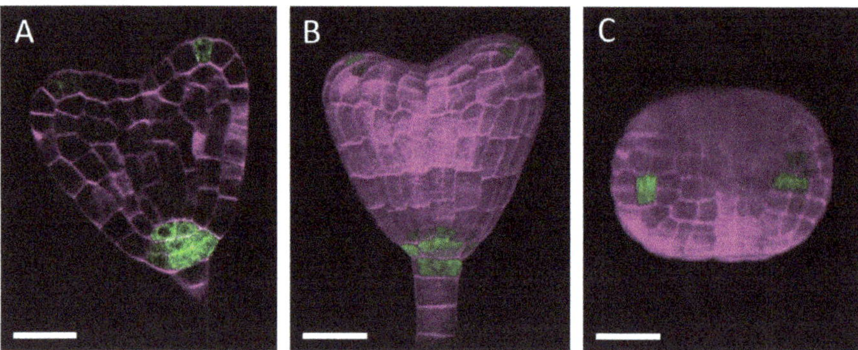

Figure 2. Confocal microscopic images of a ClearSee-processed embryo with a Green Fluorescent Protein (GFP) reporter: (**A**) frontal optical section of a heart stage embryo carrying *DR5rev::GFP* [18] and (**B,C**) 3D reconstruction of 157 serial optical sections obtained from the same embryo as in **A** in frontal (**B**) and top (**C**) views. The signals of Calcofluor White and endoplasmic reticulum (ER)-localized GFP are represented in magenta and green, respectively. Bars = 20 µm.

We next applied our protocol to other reporter lines that express different fluorescent proteins with different cellular localizations from those of *DR5rev::GFP*. The embryos of *pARF5-n3GFP*, which produce a three-tandem repeat of GFP (3xGFP) with the SV40 nuclear localization signal (NLS) [23], give clear signals in regions including provascular tissues at stages from the heart to bending-cotyledon stages (Figure 3A–C), showing that our protocol is also applicable to a nuclear-localized version of the fluorescent protein accumulated in inner tissues. We also tested samples simultaneously expressing GFP and Red Fluorescent Protein (RFP). When we examined embryos producing WUSCHEL (WUS)-3xGFP under the native regulatory sequences of the *WUS* gene (*gWUS-GFP3* [24]) and HISTONE 2B (H2B) fused to mScarlet (an RFP derivative [25]) under the control of *CLAVATA3* (*CLV3*) regulatory sequences [26] (*pCLV3:H2B-mScarlet*), green signals in the L3 layer as well as red signals in the L1–L3 layers of the shoot apex was clearly detectable (Figure 3D). These results show that our modified ClearSee protocol enables visualization of different fluorescent proteins (GFP and RFP) with different tags (ER retention, SV40-NLS, WUS, and H2B) at a range of embryonic stages (heat to bending-cotyledon stages).

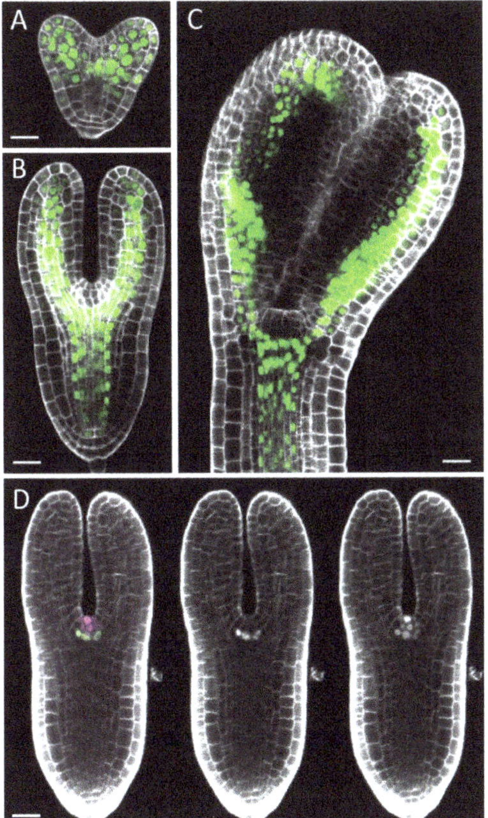

Figure 3. Confocal microscopic images of ClearSee-processed embryos with various fluorescent reporters: (**A–C**) frontal optical sections of heart (**A**), torpedo (**B**), and bent-cotyledon (**C**) stage embryos of *pARF5-n3GFP* and (**D**) a frontal optical section of the *gWUS-GFP3 pCLV3:H2B-mScarlet* embryo, displaying signals of the two fluorescent proteins (left), GFP alone (middle), and mScarlet alone (right) together with Calcofluor White signals. In (**A–C**) and the left panel of **D**, the signals of Calcofluor White, GFP, and mScarlet are represented in grey scale, green, and magenta, respectively. In the middle and right panels of **D**, the signals of Calcofluor white, GFP, and mScarlet are all represented in grey scale. Bars = 20 μm.

3. Materials and Equipment

3.1. Plant Materials

The *Arabidopsis thaliana DRrev::GFP* and *pARF5-n3GFP* reporter lines were obtained from the Arabidopsis Biological Resource Center (ABRC stock number CS9361 and CS67076, respectively). To construct *pCLV3::H2B-mScarlet*, the *Arabidopsis thaliana H2B* (AT5G22880) coding region was PCR amplified from the wild-type Col-0 genomic DNA using the primer set H2B_F and H2B_R. An additional round of PCR was performed using the attB1 and attB2 primers. The PCR product was then cloned into pDONR221 vector using BP clonase II (Thermo Fischer Scientific, Waltham, USA). The entry clone harboring H2B was PCR amplified using the primer set H2B_entry_F and H2B_entry_R, and the mScarlet coding sequence was PCR amplified from the pmScarlet_C1 [25] (Addgene #85042, Watertown, USA) using the primer set mScarlet_F and mScarlet_R. These two PCR products were integrated using In-Fusion HD Cloning Kit (TaKaRa, Kusatsu, Japan) to give *H2B-mScarlet pDONR221*. In parallel, the SalI-SacI fragment of pBU14 containing 5′ and 3′ regulatory sequences of the *CLV3* gene [26] was transferred to the corresponding sites of pBIN41, a pBIN19-derived binary vector carrying a hygromycin resistance gene, yielding *CLV3p-CLV3t pBIN41*. The *H2B-mScarlet* sequence was amplified from *H2B-mScarlet pDONR221* using the primer set H2B_mScarlet_0343_F and H2B_mScarlet_0343_R and was cloned into the BamHI site of *CLV3p-CLV3t pBIN41* by using NEBuilder HiFi DNA Assembly Master Mix (New England Biolab, Ipswich, USA) to yield *pCLV3::H2B-mScarlet*, which was transformed to plants carrying both *gWUS-GFP3* [24] and *RPS5Ap:5mCUC1-GR* [27]. The sequences of the primers are listed in Supplementary Table S1. Plants were grown as described previously [27]. Seeds were surface-sterilized using 10% commercial bleach (Kao Corporation, Tokyo, Japan) and were sown on plates containing half-strength Murashige-Skoog salts, 1% sucrose, and 0.5% gellan gum (Fujifilm Wako Pure Chemical Cooperation, Osaka, Japan). After incubation for 4–7 days at 4 °C in the dark, the plates were incubated in a growth chamber at 23 °C under constant white light. After incubation for 2 weeks, the plants were transplanted onto soil and grown under constant white light or a cycle of 8 h dark/16 h light.

3.2. Solutions

The solutions, 7% glucose in water (w/v), fixative (4% paraformaldehyde in phosphate buffered saline (PBS, w/v)), washing buffer (PBS; 130 mM NaCl, 7 mM Na_2HPO_4, and 3 mM NaH_2PO_4; pH 7.0), ClearSee (10% xylitol (w/v), 15% sodium deoxycholate (w/v), and 25% urea (w/v) in water), and staining solution (100 µg/mL Calcofluor White M2R (F3543, Sigma-Aldrich, St. Louis, MO, USA) in ClearSee) were prepared as described previously [11]. For preparing ClearSee, the reagents were first dissolved in the 0.6 volume of water and then the final volume was adjusted by adding extra water.

3.3. Equipment

Equipment consisted of stereo microscopes equipped with a transmitted light unit capable of oblique illumination (Stemi 2000-C and Stemi 508 Stand KLAB, Carl Zeiss, Oberkochen, Germany; SMZ1270, Nikon, Tokyo, Japan), the confocal microscope (Leica TCS-SPE, Leica microsystems GmbH, Wetzlar, Germany; Olympus FV3000, Tokyo, Japan), forceps (Dumont #5, Manufactures D'Outils Dumont SA, Montignez, Switzerland), and dishes (IWAKI non-treated 35-mm culture dishes 1000-035, IWAKI, Shizuoka, Japan). For Z-stack image acquisition of confocal microscopy, a 63× oil-immersion and a 40× dry objective lenses were used. Calcofluor White and GFP were excited by 405 nm and 488 nm laser lines, respectively, and were detected using 410–480 nm and 490–540 nm filter settings, respectively, with the sequential line scan mode. Images were processed using Leica LAS-X software (Leica microsystems GmbH, Wetzlar, Germany) and ImageJ [21].

Supplementary Materials: The following are available online at https://www.mdpi.com/2223-7747/10/2/190/s1, Movie S1: Z-stack images of 157 serial optical sections used for Figure 2; Table S1: Primers used in this study.

Author Contributions: Conceptualization, M.A.; methodology, M.A.; validation, A.I. and M.A.; formal analysis, A.I.; investigation, A.I., M.Y., T.S., A.O., and M.A.; resources, T.I., S.S., and M.A.; data curation, M.A.; writing—original draft preparation, M.A.; writing—review and editing, A.I., T.I., S.S., and M.A.; visualization, A.I. and M.A.; project administration, M.A.; funding acquisition, M.A. All authors have read and agreed to the published version of the manuscript.

Funding: This work was supported by the Ministry of Education, Culture, Sports, Science, and Technology of Japan (grant No. 24114009, 18H04842, and 20H04889 to M.A.); the Japan Society for the Promotion of Science (grant No. 16K07401 to M.A.); Takeda Science Foundation (to M.A.); and International Research Organization for Advanced Science and Technology (to T.I., M.A.).

Acknowledgments: We thank Arabidopsis Biological Resource Center for the *DR5rev::GFP* and *pARF5-n3GFP* lines, Shinobu Takada for the *gWUS-GFP3* line, Takehide Kato for pBIN41, Dorus Gadella for pmScarlet_C1, and Rüdiger Simon for pBU14. We also thank Maki Niidome, Mie Matsubara, and Kazuko Onga for technical assistance.

Conflicts of Interest: The authors declare no conflict of interest.

References

1. De Smet, I.; Beeckman, T. Asymmetric cell division in land plants and algae: The driving force for differentiation. *Nat. Rev. Mol. Cell Biol.* **2011**, *12*, 177–188. [CrossRef] [PubMed]
2. Boscá, S.; Knauer, S.; Laux, T. Embryonic development in *Arabidopsis thaliana*: From the zygote division to the shoot meristem. *Front. Plant Sci.* **2011**, *2*, 93. [CrossRef] [PubMed]
3. ten Hove, C.A.; Lu, K.J.; Weijers, D. Building a plant: Cell fate specification in the early Arabidopsis embryo. *Development* **2015**, *142*, 420–430. [CrossRef] [PubMed]
4. Scheres, B.; Wolkenfelt, H.; Willemsen, V.; Terlouw, M.; Lawson, E.; Dean, C.; Weisbeek, P. Embryonic origin of the *Arabidopsis* primary root and root meristem initials. *Development* **1994**, *120*, 2475–2487.
5. Barton, M.K.; Poethig, R.S. Formation of the shoot apical meristem in *Arabidopsis thaliana*: An analysis of development in the wild type and in the *shoot meristemless* mutant. *Development* **1993**, *119*, 823–831.
6. Mansfield, S.G.; Briarty, L.G. Early embryogenesis in *Arabidopsis thaliana*. II. The developing embryo. *Can. J. Bot.* **1991**, *69*, 461–476. [CrossRef]
7. Chakrabortty, B.; Willemsen, V.; de Zeeuw, T.; Liao, C.Y.; Weijers, D.; Mulder, B.; Scheres, B. A plausible microtubule-based mechanism for cell division orientation in plant embryogenesis. *Curr. Biol.* **2018**, *28*, 3031–3043.e3032. [CrossRef]
8. Yoshida, S.; Barbier de Reuille, P.; Lane, B.; Bassel, G.W.; Prusinkiewicz, P.; Smith, R.S.; Weijers, D. Genetic control of plant development by overriding a geometric division rule. *Dev. Cell* **2014**, *29*, 75–87. [CrossRef]
9. van Dop, M.; Liao, C.Y.; Weijers, D. Control of oriented cell division in the Arabidopsis embryo. *Curr. Opin. Plant Biol.* **2015**, *23*, 25–30. [CrossRef]
10. Bayer, M.; Slane, D.; Jürgens, G. Early plant embryogenesis—Dark ages or dark matter? *Curr. Opin. Plant Biol.* **2017**, *35*, 30–36. [CrossRef]
11. Kurihara, D.; Mizuta, Y.; Sato, Y.; Higashiyama, T. ClearSee: A rapid optical clearing reagent for whole-plant fluorescence imaging. *Development* **2015**, *142*, 4168–4179. [CrossRef] [PubMed]
12. Bougourd, S.; Marrison, J.; Haseloff, J. An aniline blue staining procedure for confocal microscopy and 3D imaging of normal and perturbed cellular phenotypes in mature *Arabidopsis* embryos. *Plant J.* **2000**, *24*, 543–550. [CrossRef] [PubMed]
13. Hasegawa, J.; Sakamoto, Y.; Nakagami, S.; Aida, M.; Sawa, S.; Matsunaga, S. Three-dimensional imaging of plant organs using a simple and rapid transparency technique. *Plant Cell Physiol.* **2016**, *57*, 462–472. [CrossRef] [PubMed]
14. Truernit, E.; Bauby, H.; Dubreucq, B.; Grandjean, O.; Runions, J.; Barthélémy, J.; Palauqui, J.C. High-resolution whole-mount imaging of three-dimensional tissue organization and gene expression enables the study of phloem development and structure in Arabidopsis. *Plant Cell* **2008**, *20*, 1494–1503. [CrossRef] [PubMed]
15. Vijayan, A.; Tofanelli, R.; Strauss, S.; Cerrone, L.; Wolny, A.; Strohmeier, J.; Kreshuk, A.; Hamprecht, F.A.; Smith, R.S.; Schneitz, K. A digital 3D reference atlas reveals cellular growth patterns shaping the *Arabidopsis* ovule. *eLife* **2021**, *10*. [CrossRef]
16. Ueda, M.; Kimata, Y.; Kurihara, D. Live-cell imaging of zygotic intracellular structures and early embryo pattern formation in *Arabidopsis thaliana*. In *Plant Embryogenesis: Methods and Protocols*; Bayer, M., Ed.; Springer: New York, NY, USA, 2020; pp. 37–47.
17. Hughes, R. Determinants of Seed Size and Yield in *Arabidopsis thaliana*. Ph.D. Thesis, University of Bath, Bath, UK, 2009.
18. Wilson, E.E.; Chambers, W.; Pelc, R.; Nothnagle, P.; Davidson, M.W. Stereomicroscopy in neuroanatomy. In *Neurohistology and Imaging Techniques*; Pelc, R., Walz, W., Doucette, J.R., Eds.; Springer: New York, NY, USA, 2020; pp. 245–274.
19. Friml, J.; Vieten, A.; Sauer, M.; Weijers, D.; Schwarz, H.; Hamann, T.; Offringa, R.; Jürgens, G. Efflux-dependent auxin gradients establish the apical-basal axis of *Arabidopsis*. *Nature* **2003**, *426*, 147–153. [CrossRef]
20. Benkova, E.; Michniewicz, M.; Sauer, M.; Teichmann, T.; Seifertova, D.; Jürgens, G.; Friml, J. Local, efflux-dependent auxin gradients as a common module for plant organ formation. *Cell* **2003**, *115*, 591–602. [CrossRef]
21. Schindelin, J.; Arganda-Carreras, I.; Frise, E.; Kaynig, V.; Longair, M.; Pietzsch, T.; Preibisch, S.; Rueden, C.; Saalfeld, S.; Schmid, B.; et al. Fiji: An open-source platform for biological-image analysis. *Nat. Methods* **2012**, *9*, 676–682. [CrossRef]
22. Barbier de Reuille, P.; Routier-Kierzkowska, A.L.; Kierzkowski, D.; Bassel, G.W.; Schüpbach, T.; Tauriello, G.; Bajpai, N.; Strauss, S.; Weber, A.; Kiss, A.; et al. MorphoGraphX: A platform for quantifying morphogenesis in 4D. *eLife* **2015**, *4*, 05864. [CrossRef]
23. Rademacher, E.H.; Möller, B.; Lokerse, A.S.; Llavata-Peris, C.I.; van den Berg, W.; Weijers, D. A cellular expression map of the Arabidopsis *AUXIN RESPONSE FACTOR* gene family. *Plant J.* **2011**, *68*, 597–606. [CrossRef]
24. Tucker, M.R.; Hinze, A.; Tucker, E.J.; Takada, S.; Jürgens, G.; Laux, T. Vascular signalling mediated by ZWILLE potentiates WUSCHEL function during shoot meristem stem cell development in the *Arabidopsis* embryo. *Development* **2008**, *135*, 2839–2843. [CrossRef] [PubMed]
25. Bindels, D.S.; Haarbosch, L.; van Weeren, L.; Postma, M.; Wiese, K.E.; Mastop, M.; Aumonier, S.; Gotthard, G.; Royant, A.; Hink, M.A.; et al. mScarlet: A bright monomeric red fluorescent protein for cellular imaging. *Nat. Methods* **2017**, *14*, 53–56. [CrossRef] [PubMed]
26. Brand, U.; Grünewald, M.; Hobe, M.; Simon, R. Regulation of *CLV3* expression by two homeobox genes in Arabidopsis. *Plant Physiol.* **2002**, *129*, 565–575. [CrossRef] [PubMed]
27. Takeda, S.; Hanano, K.; Kariya, A.; Shimizu, S.; Zhao, L.; Matsui, M.; Tasaka, M.; Aida, M. CUP-SHAPED COTYLEDON1 transcription factor activates the expression of *LSH4* and *LSH3*, two members of the ALOG gene family, in shoot organ boundary cells. *Plant J.* **2011**, *66*, 1066–1077. [CrossRef] [PubMed]

 *plants*

Technical Note

Detection of Embryonic Suspensor Cell Death by Whole-Mount TUNEL Assay in Tobacco

Ce Shi, Pan Luo, Peng Zhao * and Meng-Xiang Sun

State Key Laboratory of Hybrid Rice, College of Life Sciences, Wuhan University, Wuhan 430072, China; shice@whu.edu.cn (C.S.); panluo@whu.edu.cn (P.L.); mxsun@whu.edu.cn (M.-X.S.)
* Correspondence: pzhao2000@whu.edu.cn; Tel.: +027-38752378

Received: 3 July 2020; Accepted: 10 September 2020; Published: 12 September 2020

Abstract: Embryonic suspensor in angiosperms is a short-lived structure that connects the embryo to surrounding maternal tissues, which is necessary for early embryogenesis. Timely degeneration via programed cell death is the most distinct feature of the suspensor during embryogenesis. Therefore, the molecular mechanism regulating suspensor cell death is worth in-depth study for embryonic development. However, this process can hardly be detected using conventional methods since early embryos are deeply embedded in the seed coats and inaccessible through traditional tissue section. Hence, it is necessary to develop a reliable protocol for terminal deoxynucleotidyl transferase (TdT) dUTP Nick-End Labeling (TUNEL) analysis using limited living early embryos. Here, we provide a detailed protocol for the whole-mount detection of suspensor cell death using a TUNEL system in tobacco. This method is especially useful for the direct and rapid detection of the spatial-temporal characters of programed cell death during embryogenesis, as well as for the diminishment of the artifacts during material treatment by traditional methods.

Keywords: tobacco; embryogenesis; suspensor; programmed cell death; TUNEL

---

1. Introduction

The suspensor is a terminally differentiated embryonic structure, which connects the embryo to surrounding endosperms and seed coats in plants and is necessary for embryonic development by transporting nutrients and hormones from the mother tissues to the embryo [1–3]. A well-known characteristic of the suspensor is the timely initiation of programmed cell death (PCD) [4,5]. During this process, some classic markers of eukaryotic PCD have been observed in suspensor PCD, such as DNA fragmentation, nuclear degradation, and caspase-like activities [4,6–12]. Therefore, suspensor has been considered as an ideal model to investigate the molecular mechanism of PCD in plant development [2]. However, because embryos are deeply embedded in the maternal tissues, it is difficult to observe the spatial and temporal dynamics of suspensor PCD directly by conventional methods. Although the methods for the detection of suspensor PCD have been established for years in a few plants with a huge suspensor structure, such as *Picea abies* [6–9], *Vicia faba* [10], and *Phaseolus coccineus* [11,12]. As the dynamic in situ signals of PCD at the single cell level become more and more important, the establishment of a suitable technique to meet these requirements is obviously needed.

It was previously described that terminal deoxynucleotidyl transferase (TdT) dUTP Nick-End Labeling (TUNEL) is an assay to detect broken DNA fragmentation in situ [13]. This method depends on the template-independent identification of blunt ends of double stranded DNA breaks by TdT. Then, the enzyme catalyzes the addition of fluorescein labeled nucleotides to the 3′-hydroxyl termini of DNA ends, which can be visualized by fluorescence microscopy [14]. For example, to investigate integument tapetum PCD in tobacco, this tissue-specific PCD has been detected by sectioning-based TUNEL assay [15]. During the past decade, we have discussed a series of works about suspensor

PCD in tobacco [3,16,17]. Combined with our isolation technique of living early embryos [18], here, we describe a detailed protocol for the whole-mount detection of suspensor PCD using a TUNEL system. Due to its visualization and convenience, this method will be not only widely applied in the determination of the spatial-temporal characters of suspensor PCD during whole process of embryonic development in plants; it also will be helpful for detecting the embryonic cell viability in mutants with abortive embryos.

2. Results

2.1. Preparation of Hand-Made Tools

Isolation of embryos is helpful for direct observing the initiation of suspensor PCD. To date, isolation of tiny early embryos still remains technically challenging. Only a few studies reported the methods for the isolation of early embryos using either laser-capture microdissection (LCM) [19] or manual isolation [19,20]. However, the LCM equipment is not commonly available, and not suitable for isolating living early embryos. To establish a convenient and reliable protocol for isolating early embryos, we developed a set of hand-made tools for the micromanipulation; see Figure 1A–C. Among them, hand-made glass needles (Figure 1B) and the hand-made capillary pipette with latex tubing (Figure 1C) are critical for embryo isolation. Here, we describe the manual preparation of these key tools in detail.

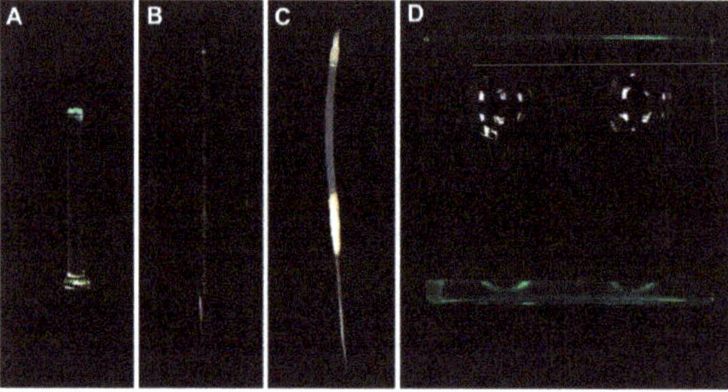

Figure 1. Tools used for the isolation of living tobacco embryos and the terminal deoxynucleotidyl transferase (TdT) dUTP Nick-End Labeling (TUNEL) assay. (**A**) A hand-made glass pestle; (**B**) a hand-made glass needle; (**C**) a hand-made capillary pipette sealed with the latex tubing; (**D**) a thick glass slide with double well concavity.

Firstly, hold one end of the glass tube and clamp the other end with tweezers (Figure 2A,B). Secondly, according the different length, place the glass tube on the flame and quickly pull the tweezers horizontally to make the glass tube form a thinner section (Figure 2C), which is the key step and requires trial and error. Then, cut the glass tube carefully with an emery wheel to make the hand-made glass needles (Figure 2D,E). To assemble the hand-made capillary pipette with a latex tubing, we prepare a 4 cm of flexible latex tube and a 1 cm of sealed glass tube (Figure 2F). Check the integrity and diameter of the glass nozzle under a microscope (Figure 2G,H); then choose the microcapillary tips with a diameter of around 200 μm (Figure 2G) for embryo sac collection, and choose another one with a diameter of around 100 μm for embryo collection (Figure 2G). Then, insert the wide end of the glass tube into the latex tube (Figure 2I,J), and insert the sealed glass tube into the other end of the latex tube (Figure 2K,L). Finally, carefully seal the two junctions with Parafilm (Figure 1C).

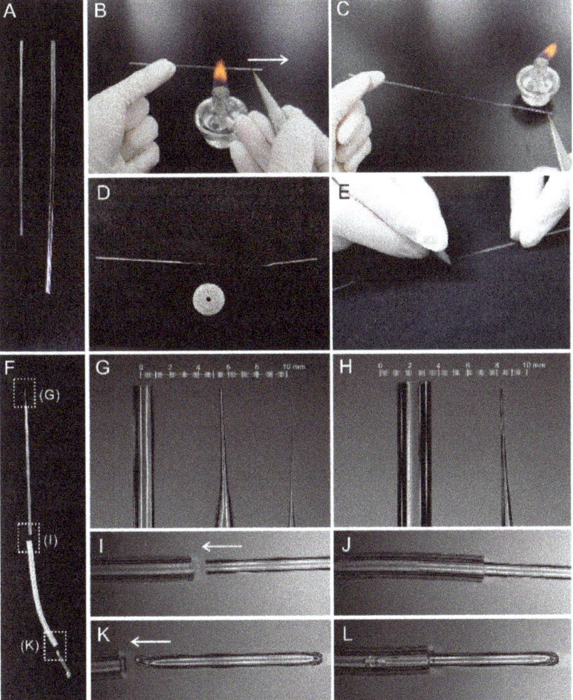

Figure 2. Manual preparation of tools for embryo isolation and TUNEL assay. (**A**) Glass microcapillary (left) and glass rod (right); (**B–E**) manual preparation of microcapillary or glass needle using a small spirit lamp; (**F**) component of a capillary pipette with latex tubing; (**G**) untreated microcapillary (left), microcapillary tips with a diameter of around 200 µm (middle), and microcapillary tips with a diameter of around 100 µm (right); (**H**) untreated glass rod (left) and fine glass needle with a diameter of around 200 µm (right); (**I–L**) assemble of a capillary pipette with latex tubing.

2.2. Collection of Living Embryos

Isolation of living embryos was performed according to the previous protocol [18]. Brief procedures are summarized in the Figure 3. In the step of embryo collection, tobacco embryos after stage 4 could be directly released from the seeds. If it is not very efficient to isolate embryo sacs by grinding the seeds, we can dissect the seeds by fine glass needles (Figure 1B) to release the embryo sacs. Usually, pressing the micropylar end gently by a fine glass needle and cutting on the seed coat by another glass needle are helpful to release the embryo sac from the seed coat. In addition, the treatment of secondary enzymolysis are required to dissect the embryos before stage 4 from the embryo sac, as previously described [18]. Wash the embryos twice with 50 µL of washing buffer, and store them in the washing buffer for subsequent TUNEL analysis.

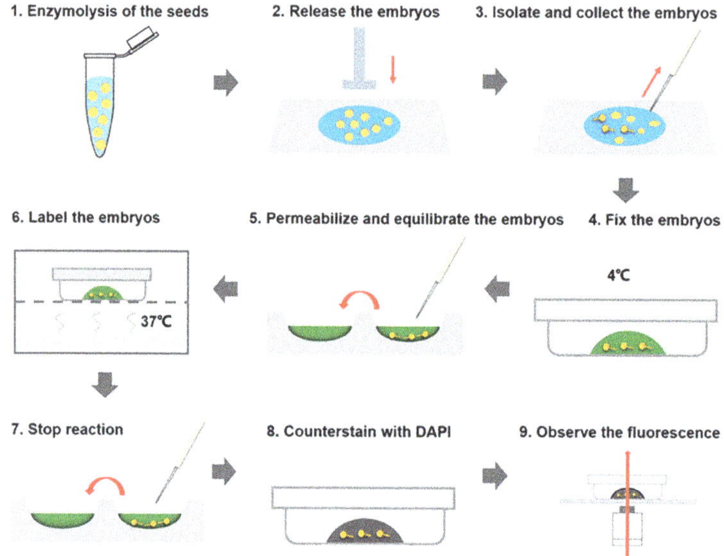

Figure 3. Schematic representation of the assay for the whole mount detection of the suspensor cell death by TUNEL. Steps 1–3 indicate the isolation of living embryos. Steps 4–9 indicate the detection of suspensor programmed cell death (PCD) by TUNEL.

2.3. TUNEL Assay

Based on the DeadEnd™ Fluorometric TUNEL System, fragmented DNA could be measured by catalytically incorporating fluorescein-12-dUTP at 3′-OH DNA ends via the Terminal Deoxynucleotidyl Transferase, Recombinant, enzyme (rTdT) [13]. The fluorescein-12-dUTP labeled DNA could be visualized directly by fluorescence microscopy.

2.3.1. Fix the Embryos

Firstly, prepare a droplet of 50 μL fresh fixation buffer in the center of a 3.5 cm Petri dish. Then, transfer these isolated embryos carefully into the fixation buffer by a hand-made capillary pipette (Figure 1C), and seal the Petri dish with Parafilm carefully. Fix the embryos for 15 min at 4 °C (Figure 3).

2.3.2. Permeabilize and Equilibrate the Embryos

During the fixation, add 200 μL of phosphate-buffered saline (PBS) solution into each well of the thick glass slide (Figure 1D). Transfer the embryos carefully into PBS solution by a hand-made capillary pipette to wash the embryos for 5 min at room temperature. Transfer these embryos carefully into another well with the fresh PBS to wash the embryos again. During the second washing, add 200 μL of PBST (PBS containing Triton® X-100) solution into a well of the thick glass slide (Figure 1D). Transfer the embryos carefully into PBST for 5 min at room temperature. After permeabiliztion in the PBST solution, wash the embryos with PBS twice. Add 100 μL of equilibration buffer into a well of the thick glass slide. Transfer the embryos carefully into the equilibration buffer, and incubate them for 5–10 min at room temperature.

2.3.3. Label the Embryos

While the embryos are incubated in the equilibration buffer, thaw the Nucleotide Mix on ice; keep the Nucleotide Mix and rTdT incubation buffer solution on the ice until use, and protect it from light. The volume of a standard reaction was enough for testing over 150 globular embryos. Then, prepare sufficient TdT reaction mix in the center of a 3.5 cm Petri dish. Transfer the embryos carefully into TdT reaction solution, and carefully seal the Petri dish with Parafilm. Incubate the Petri dish in a humidified chamber for 60 min at 37 °C, and avoid exposure to light from this step forward (Figure 1).

Prepare a negative control incubation buffer without rTdT by combining 45 μL of equilibration Buffer, 5 μL of Nucleotide Mix and 1 μL of ddH$_2$O. This step is optional because the unspecific background could hardly be detected.

If a positive control is desired, treat the embryos with DNase I as the following procedure. Add 100 μL of DNase I buffer to the fixed embryos, and incubate them for 5 min at room temperature. Transfer the embryos into 100 μL of DNase I buffer containing 10 units/mL of DNase I, and incubate them for 10 min at room temperature. After DNase I treatment, wash the embryos with the PBS solution twice.

During the labeling reaction, add 200 μL of 2 × SSC (Saline-sodium citrate) solution into a well of the thick glass slide. Transfer the embryos carefully in 200 μL of 2 × SSC solution to stop the reaction for 15 min at room temperature. Wash the embryos with the PBS solution twice to remove unincorporated fluorescein-12-dUTP. Stain the embryos in 40 μL of 1 × 4′,6-diamidino-2-phenylindole (DAPI) solution in the dark for 5 min at room temperature. Wash the embryos in the PBS solution twice.

2.3.4. Analyze the Fluorescence

The samples were then observed under a confocal microscope, with the following parameter settings: DAPI (λ_{ex} 364 nm; λ_{em} 460 ± 20 nm) and fluorescein (λ_{ex} 488 nm; λ_{em} 520 ± 20 nm) (Figure 4). If the embryos have been labeled with other fluorescence proteins [e.g., GFP (Green fluorescent protein) or YFP (Yellow fluorescent protein)], we suggest to detect the TUNEL signal using In Situ Cell Death Detection Kit, TMR (Tetramethylrhodamine) red (Roche), TUNEL (λ_{ex} 554 nm; λ_{em} 580 ± 20 nm). The protocol is almost the same as mentioned above; see the manufacturer's manual for more details.

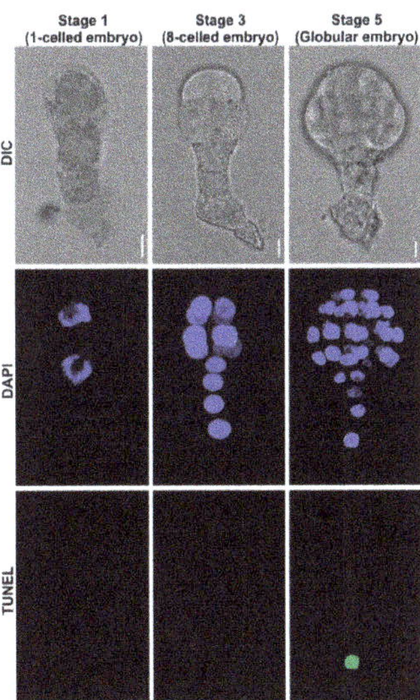

Figure 4. Tobacco embryonic suspensor PCD analyzed using the present protocol. PCD in early embryos at different developmental stages were analyzed. The blue channel indicated embryonic cell nucleus stained with DAPI (4′,6-diamidino-2-phenylindole). The green channel indicated TUNEL-positive suspensor cell. Bar = 10 μm.

3. Materials and Equipment

3.1. Plant Materials

Nicotiana tabacum var. SR1 plants were grown on natural soil in the greenhouse under a 16 h light/8 h dark photoperiod at 25 °C.

3.2. Reagents

NaCl (10019318; Sinopharm Chemical Reagent Co. Ltd., Shanghai, China), KCl (10016318; Sinopharm Chemical Reagent Co. Ltd., Shanghai, China), Na_2HPO_4 (10020318; Sinopharm Chemical Reagent Co. Ltd., Shanghai, China), KH_2PO_4 (10017618; Sinopharm Chemical Reagent Co. Ltd., Shanghai, China), Sodium citrate (W302600; Sigma, St. Louis, MO, USA), D-Mannitol (M4125; Sigma, USA), Cellulase R-10 (Yakult Pharmaceutical Industry Co. Ltd., Tokyo, Japan), Macerozyme R-10 (Yakult Pharmaceutical Industry Co. Ltd., Japan), Mineral oil (M5904; Sigma, USA), 2-(N-Morpholino) ethanesulfonic acid hydrate (MES) (M8250; Sigma, USA), Paraformaldehyde (158127; Sigma, USA), Triton® X-100 (0694; Amresco, Solon, OH, USA), DAPI (D9542; Sigma, USA), DeadEnd™ Fluorometric TUNEL System (G3250; Promega, Madison, MI, USA), RQ1 RNase-Free DNase (M6101; Promega, USA), and In Situ Cell Death Detection Kit, TMR red (12156792910; Roche, Basel, Switzerland) (if necessary).

3.3. Equipment

Inverted microscope (CK2; OLYMPUS, Tokyo, Japan), Confocal microscope (SP8; Leica, Wetzlar, Germany), fine tweezer, glass microscope slide (80312; CITOTEST, Haimen, China), Glass microcapillary (2177401; HIRSCHMANN, Eberstadt, Germany), fine glass rod (2.5 mm × 130 mm, custom-made; YUNCHENG, China), flexible latex tube (inner diameter = 1.5 ± 1 mm, outer diameter = 2.3 ± 1 mm; DAOGUAN, Shanghai, China), emery wheel (3.2 mm × 21 mm; JIAKANG, China), parafilm M (PM-996; Bemis, Neenah, WI, USA), hand-made glass pestle (Figure 1A), hand-made glass needles (Figure 1B), hand-made capillary pipette with latex tubing (Figure 1C), Petri dishes for microscope observations (3.5 cm), thick glass slide with double well concavity (5-mm, custom-made; HUICHENG, Taizhou, China) (Figure 1D), humidified chamber (a light-proof box with moisture gauze to keep wet), and incubator (LRH-400A; RUIHUA, Wuhan, China).

3.4. Solutions

Washing buffer: 13% D-mannitol, 0.058% MES, pH 5.8. Enzyme buffer I: 1% Cellulase R-10 and 0.8% Macerozyme R-10 dissolved in the washing buffer, pH 5.8. Filter-sterilize the enzyme buffer with a 0.22-μm filter, and make single-use aliquots. Store at −20 °C. Enzyme buffer II: 0.25% Cellulase R-10 and 0.2% Macerozyme R-10 dissolved in the washing buffer, pH 5.8. Filter-sterilize the enzyme buffer with a 0.22-μm filter, and make single-use aliquots. Store at −20 °C. Phosphate-buffered saline (PBS): 137 mM NaCl, 2.7 mM KCl, 10 mM Na_2HPO_4, and 1.76 mM KH_2PO_4 in ddH_2O, pH 7.4. Fixation buffer: 4% Paraformaldehyde in PBS. Make fresh 4% paraformaldehyde in the PBS solution for each experiment. It is necessary to warm solution to dissolve paraformaldehyde at 60 °C. Preparations should be carried out in a fume hood. Store the fixation buffer at 4 °C for up to 1 week. PBST: 0.2% Triton® X-100 in PBS. Equilibration buffer (a component in the DeadEnd™ Fluorometric TUNEL System). TdT reaction solution: Add 45 μL of equilibration buffer, 5 μL of Nucleotide Mix which contains fluorescein-12-dUTP, and 1 μL of rTdT per one reaction. SSC (20×): 3 M NaCl, 0.3 M sodium citrate in ddH_2O, pH 4.5. SSC (2×): Warm 20 × SSC to room temperature to ensure that all salts are in solution. Dilute 1:10 with ddH_2O before use to generate 2 × SSC. DAPI (1000×): 1 mg DAPI in 1 mL ddH_2O. Store in the dark at 4 °C for 2 to 3 weeks. DAPI solution (1×): Dilute DAPI (1000×) with PBS before use to generate 1 × DAPI solution.

4. Discussion

Suspensor is a terminally differentiated embryonic organ, which helps the embryo to fix in the seed and transfers nutrients and plant hormones to the embryo for normal development [1,2]. Previous research demonstrated that suspensor degeneration is a kind of typical PCD. Therefore, we could study the molecular mechanism regulating plant PCD using the suspensor as a model system [2]. TUNEL assay is one of important methods to study PCD in both animals and plants. Although TUNEL assay has been applied to investigate stress induced-PCD and developmental PCD in plants for years [21–23], it is still difficult to analyze suspensor PCD using the traditional TUNEL assay methods. Based on our previous report about the isolation of living early embryos [18], here, we described a detailed method for analyzing suspensor PCD via TUNEL using limited early embryos. The equipment described here consists of an inverted microscope, glass microcapillary, fine glass rod, and flexible latex tube (Figure 1). In addition, the key hand-made tools are easy to assemble according to the introduction (Figure 2). This basic setup has been proven to be efficient and reliable in different plants [18,20]. Compared with the other available methods, this procedure offers several advantages: (i) the spatial-temporal characters of suspensor PCD could be quickly detected within 5–6 h, (ii) it can be easily adopted by other researchers due to the simple setup, (iii) it requires affordable equipment for the basic setup, and (iv) this method could also be useful for rapid detection of cell death of abortive embryos. Nevertheless, one of the main difficulties is that it requires practicing it over and over again to ensure the quick isolation and collection of living early embryos under an inverted microscope.

5. Conclusions

In conclusion, we developed a detailed protocol for detecting suspensor PCD via TUNEL. Combined with the isolation of living embryo, this method will be widely applied to investigate the spatial-temporal characters of suspensor PCD in different plants.

Author Contributions: Conceptualization C.S., M.-X.S. Methodology C.S., P.L., P.Z. Resources M.-X.S. Writing—original draft preparation C.S. Writing—review and editing M.-X.S. All authors have read and agreed to the published version of the manuscript.

Funding: This study was supported by National Natural Science Fund of China (31970340; 31630094) and the Postdoctoral Science Foundation of China (BX20200256, 2019M662570).

Conflicts of Interest: The authors declare no conflict of interest.

References

1. Kawashima, T.; Goldberg, R.B. The suspensor: Not just suspending the embryo. *Trends Plant Sci.* **2010**, *15*, 23–30. [CrossRef]

2. Peng, X.; Sun, M.-X. The suspensor as a model system to study the mechanism of cell fate specification during early embryogenesis. *Plant Reprod.* **2018**, *31*, 59–65. [CrossRef]

3. Liu, Y.; Li, X.; Zhao, J.; Tang, X.; Tian, S.; Chen, J.; Shi, C.; Wang, W.; Zhang, L.; Feng, X.; et al. Direct evidence that suspensor cells have embryogenic potential that is suppressed by the embryo proper during normal embryogenesis. *Proc. Natl. Acad. Sci. USA* **2015**, *112*, 12432. [CrossRef]

4. Zhao, P.; Zhou, X.-M.; Zhang, L.-Y.; Wang, W.; Ma, L.-G.; Yang, L.-B.; Peng, X.-B.; Bozhkov, P.V.; Sun, M.-X. A bipartite molecular module controls cell death activation in the basal cell lineage of plant embryos. *PLoS Biol.* **2013**, *11*, e1001655. [CrossRef]

5. Bozhkov, P.V.; Filonova, L.H.; Suarez, M.F. Programmed cell death in plant embryogenesis. *Curr. Top. Dev. Biol.* **2005**, *67*, 135–179. [PubMed]

6. Bozhkov, P.V.; Suarez, M.F.; Filonova, L.H.; Daniel, G.; Zamyatnin, A.A.; Rodriguez-Nieto, S.; Zhivotovsky, B.; Smertenko, A. Cysteine protease mcII-Pa executes programmed cell death during plant embryogenesis. *Proc. Natl. Acad. Sci. USA* **2005**, *102*, 14463–14468. [CrossRef] [PubMed]

7. Smertenko, A.P.; Bozhkov, P.V.; Filonova, L.H.; von Arnold, S.; Hussey, P.J. Re-organisation of the cytoskeleton during developmental programmed cell death in *Picea abies* embryos. *Plant J.* **2003**, *33*, 813–824. [CrossRef]

8.   Filonova, L.H.; Bozhkov, P.V.; Brukhin, V.B.; Daniel, G.; Arnold, S.V. Two waves of programmed cell death occur during formation and development of somatic embryos in the gymnosperm, Norway spruce. *J. Cell Sci.* **2000**, *113*, 4399–4411. [PubMed]

9.   Bozhkov, P.V.; Filonova, L.H.; Suarez, M.F.; Helmersson, A.; Smertenko, A.P.; Zhivotovsky, B.; von Arnold, S. VEIDase is a principal caspase-like activity involved in plant programmed cell death and essential for embryonic pattern formation. *Cell Death Differ.* **2004**, *11*, 175–182. [CrossRef]

10. Wredle, U.; Walles, B.; Hakman, I. DNA fragmentation and nuclear degradation during programmed cell death in the suspensor and endosperm of *Vicia faba*. *Int. J. Plant Sci.* **2001**, *162*, 1053–1063. [CrossRef]

11. Lombardi, L.; Ceccarelli, N.; Picciarelli, P.; Lorenzi, R. DNA degradation during programmed cell death in *Phaseolus coccineus* suspensor. *Plant Physiol. Biochem.* **2007**, *45*, 221–227. [CrossRef] [PubMed]

12. Lombardi, L.; Ceccarelli, N.; Picciarelli, P.; Lorenzi, R. Caspase-like proteases involvement in programmed cell death of *Phaseolus coccineus* suspensor. *Plant Sci.* **2007**, *172*, 573–578. [CrossRef]

13. Gavrieli, Y.; Sherman, Y.; Ben-Sasson, S.A. Identification of programmed cell death in situ via specific labeling of nuclear DNA fragmentation. *J. Cell Biol.* **1992**, *119*, 493–501. [CrossRef]

14. Rath, N.C.; Huff, W.E.; Bayyari, G.R.; Balog, J.M. Cell death in avian tibial dyschondroplasia. *Avian Dis.* **1998**, *42*, 72–79. [CrossRef] [PubMed]

15. Wang, W.; Xiong, H.; Lin, R.; Zhao, N.; Zhao, P.; Sun, M.-X. VPE-like protease NtTPE8 exclusively expresses in the integumentary tapetum and is involved in seed development. *J. Integr. Plant Biol.* **2018**, *61*, 598–610. [CrossRef]

16. Shi, C.; Luo, P.; Du, Y.-T.; Chen, H.; Huang, X.; Cheng, T.-H.; Luo, A.; Li, H.-J.; Yang, W.-C.; Zhao, P.; et al. Maternal control of suspensor programmed cell death via gibberellin signaling. *Nat. Commun.* **2019**, *10*, 3484. [CrossRef]

17. Luo, A.; Zhao, P.; Zhang, L.Y.; Sun, M.-X. Initiation of programmed cell death in the suspensor is predominantly regulated maternally in a tobacco hybrid. *Sci. Rep.* **2016**, *6*, 29467. [CrossRef]

18. Zhao, P.; Zhou, X.-M.; Shi, C.; Sun, M.-X. Manual isolation of living early embryos from tobacco seeds. *Methods Mol. Biol.* **2020**, *2122*, 101–111.

19. Palovaara, J.; Saiga, S.; Weijers, D. Transcriptomics approaches in the early Arabidopsis embryo. *Trends Plant Sci.* **2013**, *18*, 514–521. [CrossRef]

20. Zhou, X.-M.; Shi, C.; Zhao, P.; Sun, M.-X. Isolation of living apical and basal cell lineages of early proembryos for transcriptome analysis. *Plant Reprod.* **2019**, *32*, 105–111. [CrossRef]

21. Tripathi, A.K.; Pareek, A.; Singla-Pareek, S.L. A NAP-Family Histone chaperone functions in abiotic stress response and adaptation. *Plant Physiol.* **2016**, *171*, 2854–2868. [PubMed]

22. Biswas, M.S.; Mano, J.I. Lipid peroxide-derived short-chain carbonyls mediate hydrogen peroxide-induced and salt-induced programmed cell death in plants. *Plant Physiol.* **2015**, *168*, 885–898. [CrossRef] [PubMed]

23. Fendrych, M.; Van Hautegem, T.; Van Durme, M.; Olvera-Carrillo, Y.; Huysmans, M.; Karimi, M.; Lippens, S.; Guérin, C.J.; Krebs, M.; Schumacher, K.; et al. Programmed cell death controlled by *ANAC033/SOMBRERO* determines root cap organ size in Arabidopsis. *Curr. Biol.* **2014**, *24*, 931–940. [CrossRef] [PubMed]

Article

Effect of Paternal Genome Excess on the Developmental and Gene Expression Profiles of Polyspermic Zygotes in Rice

Ryouya Deushi [1,†], Erika Toda [1,†], Shizuka Koshimizu [2], Kentaro Yano [2] and Takashi Okamoto [1,*]

1   Department of Biological Sciences, Tokyo Metropolitan University, Tokyo 192-0392, Japan; underworld0326@gmail.com (R.D.); etoda@tmu.ac.jp (E.T.)
2   Department of Life Sciences, Meiji University, Kanagawa 214-8571, Japan; ty18004@meiji.ac.jp (S.K.); kyano@meiji.ac.jp (K.Y.)
*   Correspondence: okamoto-takashi@tmu.ac.jp
†   These authors contributed equally to this study.

Abstract: Polyploid zygotes with a paternal gamete/genome excess exhibit arrested development, whereas polyploid zygotes with a maternal excess develop normally. These observations indicate that paternal and maternal genomes synergistically influence zygote development via distinct functions. In this study, to clarify how paternal genome excess affects zygotic development, the developmental and gene expression profiles of polyspermic rice zygotes were analyzed. The results indicated that polyspermic zygotes were mostly arrested at the one-cell stage after karyogamy had completed. Through comparison of transcriptomes between polyspermic zygotes and diploid zygotes, 36 and 43 genes with up-regulated and down-regulated expression levels, respectively, were identified in the polyspermic zygotes relative to the corresponding expression in the diploid zygotes. Notably, *OsASGR-BBML1*, which encodes an AP2 transcription factor possibly involved in initiating rice zygote development, was expressed at a much lower level in the polyspermic zygotes than in the diploid zygotes.

Keywords: fertilization; male excess; parental genome; paternal genome; polyspermy; rice

Citation: Deushi, R.; Toda, E.; Koshimizu, S.; Yano, K.; Okamoto, T. Effect of Paternal Genome Excess on the Developmental and Gene Expression Profiles of Polyspermic Zygotes in Rice. *Plants* **2021**, *10*, 255. https://doi.org/10.3390/plants10020255

Academic Editor: Minako Ueda
Received: 23 December 2020
Accepted: 26 January 2021
Published: 28 January 2021

Publisher's Note: MDPI stays neutral with regard to jurisdictional claims in published maps and institutional affiliations.

1. Introduction

Fertilization is a characteristic event of eukaryotic unicellular and multicellular organisms that combines male and female genetic materials for the next generation. In the diploid zygote generated by the fusion between haploid male and female gametes, parental genomes function synergistically to ensure the faithful progression of zygotic development and the subsequent embryogenesis. In angiosperms, sporophytic generation is initiated by a double fertilization to form seeds that are consisting of three tissues, embryo, endosperm and maternal seed coat [1]. Regarding the double fertilization, one sperm cell fuses with the egg cell, resulting in the formation of a zygote, and another sperm cell fuses with the central cell to form a triploid primary endosperm cell. The zygote and primary endosperm cell respectively develop into the embryo, which carries genetic material from the parents, and the endosperm, which nourishes the developing embryo and seedling [2–4]. Of the three tissues in seeds, it has been known that the endosperm is highly sensitive to an imbalanced parental genome ratio resulting from ploidy differences between the parents [5–9].

In a recent study, the effects of parental genome imbalance on zygotic development were clarified by producing polyploid zygotes with an imbalanced parental genome ratio via the *in vitro* fertilization of isolated rice gametes and by elucidating the developmental profiles of the polyploid zygotes [10,11]. The results indicated that approximately 50%–75% of the polyploid zygotes with an excess of paternal genome content exhibited the developmental arrest, whereas most of the polyploid zygotes with an excess of maternal gamete/genome content developed normally, as diploid zygotes [10]. Notably, the paternal excess zygotes did not progress beyond the first zygotic division, although karyogamy was completed normally. These results suggest that parental genomes have different functions and are used synergistically in zygotes. Moreover, the early zygotic developmental steps, from karyogamy to the first cell division, are highly sensitive to paternal genome excess. Consistent with the possible preferential functions of parental genomes in zygotic embryogenesis, genes expressed in a monoallelic and/or parent-of-origin manner during zygotic development and/or early embryogenesis have been identified, and the functions of some monoallelic genes during early embryogenesis have been thoroughly investigated [12–18]. In addition, it has been reported that genes relating to cell cycle, RNA processing, signaling pathway and other cellular machineries are involved in zygotic division and/or development [19–26]. However, it remains unclear how parental genomes function synergistically in developing zygotes.

In the present study, we focused on the developmental characteristics of paternal excess rice zygotes (i.e., polyspermic zygotes), since the developmental arrest of the polyspermic zygote would be due to the excess male genomic content in the nucleus, wherein the imbalanced parental genomes may adversely affect zygotic development. The possible mechanism underlying the dysfunction between parental genomes is partly clarified by comparing the developmental and gene expression profiles of the polyspermic zygotes with those of diploid zygotes [10]. Therefore, development of polyspermic rice zygote was carefully monitored to identify the stage in which the developmental arrest becomes evident. Furthermore, the transcriptomes of the polyspermic zygotes and diploid zygotes were compared to determine the effects of the paternal excess on the zygote gene expression profiles.

2. Results

2.1. Developmental Profiles of Polyspermic Rice Zygotes

In this study, sperm cells isolated from transformed rice plants expressing histone H2B-GFP were used to produce zygotes for the subsequent visualization of the nucleus in developing zygotes. Diploid zygotes were produced via the electro-fusion between egg and sperm cells (Figure 1A). The zygotes developed into a two-celled embryo at 17.5 h after gamete fusion and a globular-like embryo was formed via repeated cell division at 42 h after gamete fusion (Figure 1B) [27]. Polyspermic zygotes were generated using one egg cell and two sperm cells (Figure 1C) [28]. We produced 34 polyspermic zygotes for the sequential monitoring of developmental steps from karyogamy to the first zygotic division. In an earlier study, we analyzed the developmental profiles of polyspermic zygotes daily after the gametes fused to ascertain whether the cells of the polyspermic zygotes were dividing [10], and were unsuccessful in determining exactly when the degeneration of developing polyspermic zygotes becomes apparent.

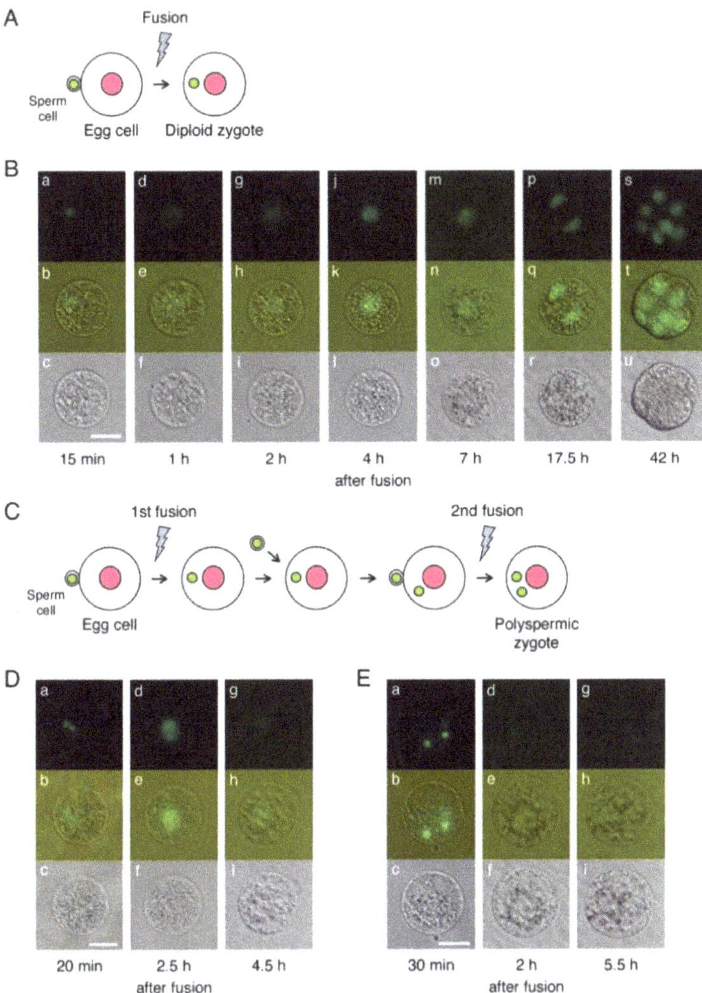

Figure 1. Developmental profiles of a diploid zygote (**A**,**B**) and polyspermic zygotes (**C–E**). (**A**) Schematic illustration of the production of diploid rice zygotes. An egg cell and a sperm cell were fused to produce a monospermic diploid zygote. (**B**) Developmental profiles of a diploid rice zygote. A sperm nucleus fluorescently labeled with H2B-GFP was observed in the zygote (a–c) and karyogamy progressed in the zygote (d–l). Thereafter, the zygote developed into a two-celled embryo (m–r) and a globular-like embryo (s–u). (**C**) Schematic illustration of the production of polyspermic rice zygotes. Two sperm cells were sequentially fused to an egg cell to produce a polyspermic zygote as described by Toda et al. (2016) [28]. (**D**) Progression of karyogamy in polyspermic zygotes. Two sperm nuclei fluorescently labeled with H2B-GFP were detected in the polyspermic zygote at 20 min after the fusion (a–c). The two nuclei then fused with an egg nucleus, resulting in a detectable zygotic nucleus (d–i). (**E**) Lack of karyogamy in polyspermic zygotes. Although two sperm nuclei fluorescently labeled with H2B-GFP were observed in the fused egg cell (a–c), the progression of karyogamy was undetectable (d–i). Pink and green circles in (**A**,**C**) indicate the egg and sperm nuclei, respectively. The gray flash symbols in (**A**,**C**) represent electro-fusions. Top, middle, and bottom panels in (**B**,**D**,**E**) represent fluorescent, merged fluorescent/bright-field, and bright-field images, respectively. Scale bars = 20 μm.

Among the 34 polyspermic zygotes, karyogamy, which involves the fusion of two sperm and one egg nuclei to form a zygotic nucleus, was detected in 30 zygotes (Figure 1D; Table 1). Karyogamy was undetectable in the other four polyspermic zygotes (Figure 1E), which subsequently degenerated. Upon the completion of karyogamy, 19 of the 30 polyspermic zygotes divided into two-celled and globular-like embryos (Figure 2A) similar to diploid zygotes (Figure 1B). Arrested development was observed in the remaining 11 polyspermic zygotes (Table 1), suggesting that approximately one-third of the polyspermic zygotes were affected by post-karyogamy defects during development. This tendency was consistent with the results of our previous analysis of the cell division profiles of polyspermic zygotes (Supplemental Table S1) [10]. The developmental profiles of the 11 polyspermic zygotes after karyogamy revealed two degeneration patterns. Specifically, for nine of the polyspermic zygotes, the cells became transparent and appeared to be highly vacuolated at approximately 11–15 h after gamete fusion (Figure 2B). Additionally, the intensity of the fluorescent signals from the H2B-GFP in the nucleus decreased to low levels (Figure 2B), and the zygotes finally degenerated. This degeneration pattern was considered to reflect the main developmental defects of polyspermic zygotes. Regarding the other two polyspermic zygotes, abnormal cellular characteristics were not evident at approximately 10–18 h after the fusion (Figure 2C), and the fluorescence intensity in the nucleus was equivalent to that of diploid and/or polyspermic zygotes which divided into two-celled embryos (Figure 1B, Figure 2A,C). However, the fluorescent signals in the nucleus of these two polyspermic zygotes became undetectable at approximately 21 h after the fusion (Figure 2C), which is just before the first zygotic division. The zygotes then degenerated without dividing (Figure 2C). These two types of degeneration profiles suggest that developmental defects can be triggered at early and late developmental stages (Figure 3), and that the early developmental stage, probably after karyogamy, is primarily when zygotic development is affected by imbalanced parental genomes. Therefore, polyspermic zygotes and diploid zygotes at 4–5 h after gamete fusion (i.e., following the completion of karyogamy) were freshly prepared for transcriptome analyses.

2.2. Gene Expression Profiles of Polyspermic Zygotes

To identify genes with misregulated expression in polyspermic zygotes, the differentially expressed genes (DEGs) between polyspermic and diploid zygotes were analyzed. Relative to the corresponding expression in the diploid zygotes, 36 and 43 genes with up-regulated and down-regulated expression levels, respectively, were identified in the polyspermic zygotes (Tables 2 and 3, Supplemental Tables S2 and S3). Expression profiles of the representative 4 up- or down-regulated genes in polyspermic zygotes were confirmed using semi-quantitative RT-PCR (Figure 4). The enriched gene ontology (GO) terms among the up-regulated genes in the polyspermic zygotes were related to chromatin/chromosomal assembly/organization (Supplemental Table S4). Whereas, no GO term was enriched among the down-regulated genes.

Table 1. Developmental profiles of diploid and polyspermic rice zygotes.

Ploidy	Gametes Used for Fusion	No. of Zygotes Produced	No. of Zygotes That Developed to Specific Growth Stages			
			Karyogamy	Two-Celled Embryo	Globular-Like Embryo	Cell Mass
2X	Egg + Sperm	22	18	18	18	18
3X	Egg + Sperm + Sperm	34	30	19	17	17

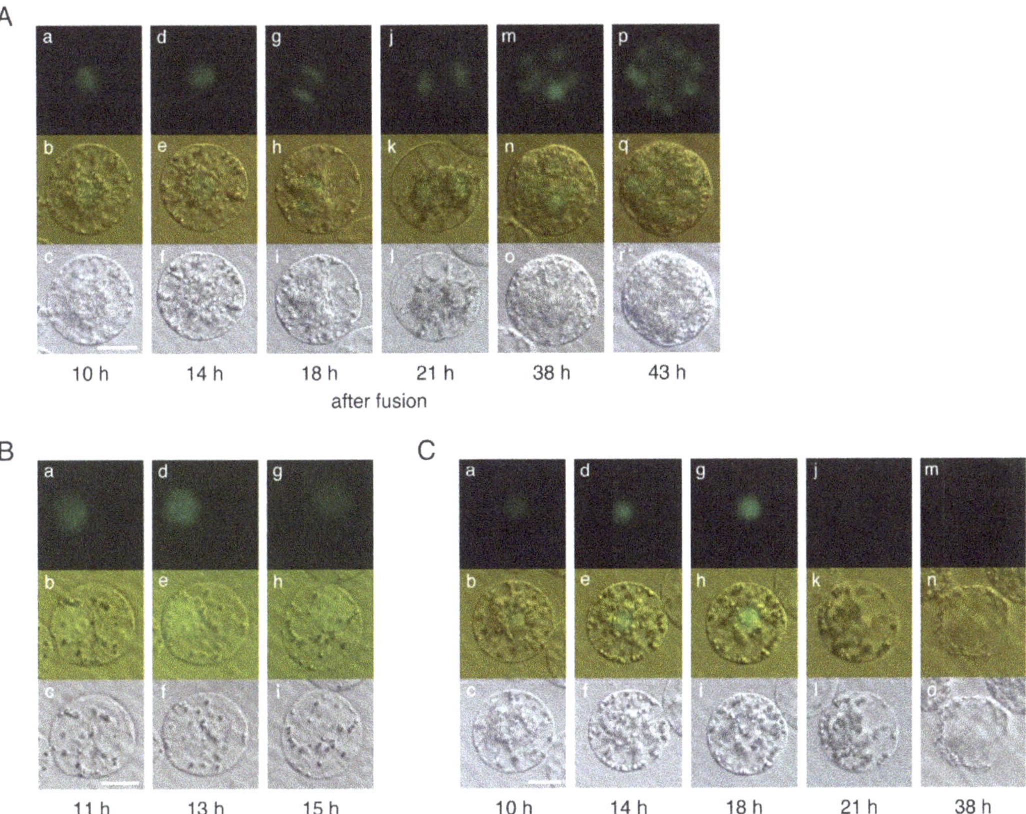

Figure 2. Developmental profiles of polyspermic rice zygotes after karyogamy. An egg cell was serially fused with two sperm cells expressing H2B-GFP, and the resulting zygote was analyzed. (**A**) After karyogamy, the polyspermic zygotes developed and divided into a two-celled embryo (a–l) and a globular-like embryo (m–r). (**B**) Developmental arrest of polyspermic zygotes (pattern I). Although the H2B-GFP signal was detectable in the zygotic nucleus, zygotes were highly vacuolated and became transparent (a–i) before they degenerated. (**C**) Developmental arrest of polyspermic zygotes (pattern II). The H2B-GFP signal was clearly detected in the zygotic nucleus during development (a–i); however, the fluorescent signal decreased and was undetectable at approximately 21 h after the fusion (j–l). The zygotes degenerated without dividing (m–o). Top, middle, and bottom panels represent fluorescent, merged fluorescent/bright-field, and bright-field images, respectively. Scale bars = 20 μm.

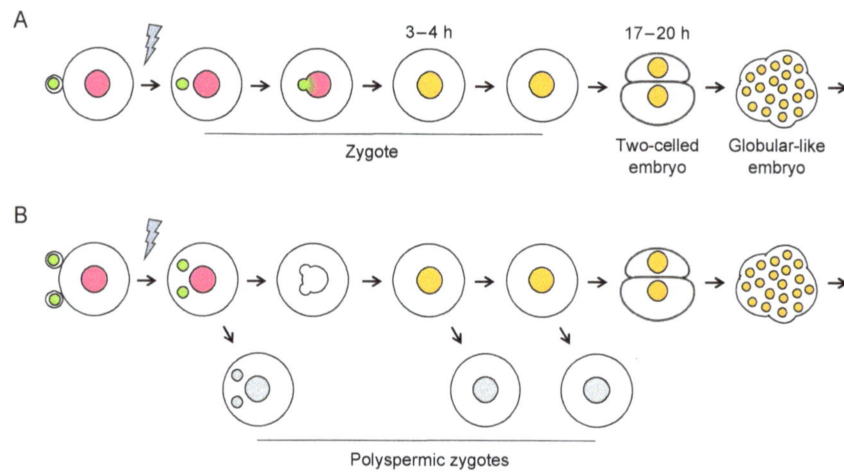

Figure 3. Schematic diagram of the early development of the diploid zygote (**A**) and polyspermic zygote (**B**). The times required for the completion of karyogamy (ca. 3–4 h) and the first cell division (ca. 17–20 h) are provided. Pink, green, and orange circles indicate the egg, sperm, and zygotic nuclei, respectively. Gray circles indicate the egg, sperm, and zygotic nuclei in the polyspermic zygotes that exhibited arrested development. The gray flash symbols represent electro-fusions.

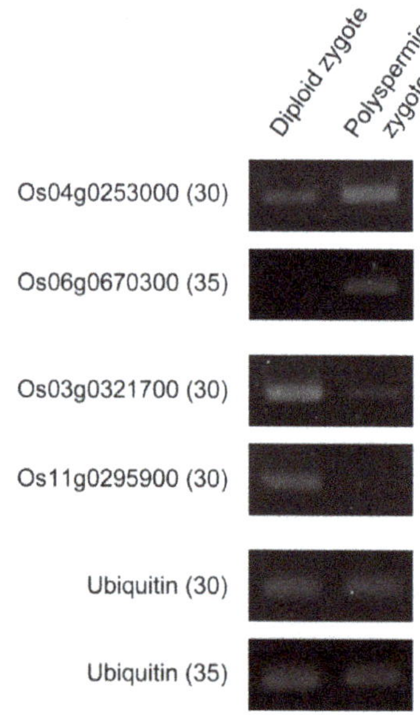

Figure 4. Expression patterns of 4 genes whose expression levels were putatively up- or down-regulated in polyspermic zygotes. Semi-quantitative RT-PCR was performed on cDNAs synthesized from diploid and polyspermic zygotes using specific primers for the putatively up-regulated genes, Os04g0253000 and Os06g0670300 (Table 2) and down-regulated genes, Os03g0321700 and Os11g0295900 (Table 3) in polyspermic zygotes. Ubiquitin cDNA was used as an internal control. Numbers in parentheses indicate the number of PCR cycles. Primer sequences are presented in Supplementary Table S5.

Table 2. Identified genes whose expression levels were putatively up-regulated in polyspermic zygote.

Gene ID	Expression Level (Averaged TPM)		p Value	q Value	RAP-DB Description
	Diploid Zygotes	Poly-Spermic Zygotes			
Os01g0115600	0.0	99.8	2.17×10^{-5}	0.01900855	Similar to LRK14
Os01g0136000	0.0	461.4	6.81×10^{-8}	2.03×10^{-4}	Similar to cytosolic class I small heat-shock protein HSP17.5
Os01g0149900	0.2	199.0	6.03×10^{-6}	0.00748057	Conserved hypothetical protein
Os01g0612500	0.2	430.2	2.16×10^{-5}	0.01900855	Phospholipase/carboxylesterase domain containing protein
Os01g0760000	0.2	137.8	5.66×10^{-6}	0.00743199	Similar to dynein light chain
Os01g0778500	0.0	194.2	1.56×10^{-6}	0.00268023	Similar to predicted protein
Os01g0794800	2.5	304.5	4.17×10^{-6}	0.00564972	Similar to subtilase
Os01g0866200	8033.4	28317.5	1.08×10^{-5}	0.01179179	Similar to histone H3
Os02g0773200	98.0	3120.0	1.39×10^{-5}	0.01409845	UspA domain containing protein
Os03g0119900	5982.5	22237.0	6.93×10^{-6}	0.00836748	Similar to histone H4
Os03g0217900	1420.4	14539.4	7.16×10^{-9}	3.22×10^{-5}	Hypothetical protein
Os03g0227800	2108.4	9163.3	3.81×10^{-6}	0.00549041	Conserved hypothetical protein
Os03g0292100	384.8	1346.8	8.05×10^{-5}	0.04732313	Hypothetical conserved gene
Os03g0670700	756.9	3743.1	4.96×10^{-5}	0.03357786	Similar to glycine rich RNA binding protein
Os03g0675600	0.2	404.2	9.73×10^{-6}	0.01086596	Similar to phytosulfokines 3 precursor
Os04g0253000	461.5	6098.4	7.25×10^{-5}	0.04321752	Similar to histone H1
Os04g0565500	0.0	263.3	2.93×10^{-7}	6.88×10^{-4}	Similar to OSIGBa0158F05.8 protein
Os04g0668800	387.1	2053.6	2.68×10^{-6}	0.00443244	Putative thiol-disulphide oxidoreductase DCC family protein
Os05g0152201	1501.4	8510.0	5.84×10^{-6}	0.00745909	Conserved hypothetical protein
Os05g0475400	0.0	164.1	1.90×10^{-5}	0.01728341	Similar to alanine:glyoxylate aminotransferase-like protein
Os06g0597250	7031.8	18692.8	4.36×10^{-5}	0.03172249	Similar to B protein
Os06g0670300	0.0	140.8	5.75×10^{-5}	0.03626026	MYB-like transcription factor
Os07g0483500	0.0	179.3	2.29×10^{-5}	0.01934667	Similar to phosphoribosyltransferase
Os08g0388300	0.2	123.7	7.68×10^{-6}	0.00902684	NB-ARC domain containing protein
Os08g0409900	0.2	156.2	1.63×10^{-5}	0.01618585	Major facilitator superfamily protein
Os09g0411500	1295.0	3925.1	8.55×10^{-5}	0.04963598	Similar to predicted protein
Os09g0433600	1651.2	5276.6	5.30×10^{-5}	0.03536701	Similar to histone H4
Os09g0457100	0.8	184.3	4.83×10^{-5}	0.03323256	Cytochrome P450 family protein
Os09g0483400	560.5	8011.9	1.27×10^{-6}	0.00227203	Similar to ubiquitin/ribosomal fusion protein
Os09g0551600	8147.5	19900.5	5.76×10^{-5}	0.03626026	Similar to HMGd1 protein
Os10g0539500	8203.4	28218.2	1.23×10^{-6}	0.00227203	Similar to histone H4
Os11g0222800	0.0	205.4	6.49×10^{-5}	0.04028685	Similar to LGC1
Os11g0533400	0.0	426.6	5.85×10^{-11}	6.54×10^{-7}	Conserved hypothetical protein
Os11g0550100	0.0	317.8	3.74×10^{-7}	8.36×10^{-4}	Similar to NB-ARC domain containing protein
Os12g0127200	3.2	249.1	3.96×10^{-5}	0.02999367	Harpin-induced 1 domain containing protein
Os12g0438000	2594.6	8152.9	1.30×10^{-5}	0.01384279	Similar to histone H2A

Table 3. Identified genes with putatively down-regulated expression levels in polyspermic zygotes.

| Gene ID | Expression Level (Averaged TPM) | | p Value | q Value | RAP-DB Description |
	Diploid Zygotes	Poly-Spermic Zygotes			
Os01g0341200	612.4	2.2	3.77×10^{-6}	0.005490414	Tubulin, conserved site domain containing protein
Os01g0611900	107.11	0	2.24×10^{-5}	0.019231324	Pentatricopeptide repeat domain containing protein
Os01g0622033	368.254	1.778	8.79×10^{-5}	0.049729216	Hypothetical gene
Os01g0737700	415.054	0.6	7.21×10^{-9}	3.22×10^{-5}	Similar to OSIGBa0101A01.4 protein
Os01g0804200	314.4	0.2	4.58×10^{-9}	2.93×10^{-5}	Cytochrome P450 of the CYP94 subfamily, response to wounding and salt stress
Os01g0931400	719.77	60.236	8.76×10^{-5}	0.049729216	Thiamin pyrophosphokinase, eukaryotic domain containing protein
Os02g0281200	701.436	0.6	4.58×10^{-12}	6.82×10^{-8}	Similar to NBS-LRR protein
Os02g0483500	3362.754	455.8	8.09×10^{-7}	0.001643243	Transferase family protein
Os02g0755900	194.4	0.2	4.46×10^{-5}	0.031722495	UDP-glucuronosyl/UDP-glucosyltransferase domain containing protein
Os02g0812000	208.764	0.2	5.71×10^{-5}	0.036260256	NAD(P)-binding domain containing protein
Os03g0299800	197.17	0	1.39×10^{-5}	0.014098454	Protein of unknown function Cys-rich family protein
Os03g0321700	2104.428	182.582	7.17×10^{-5}	0.043217519	Similar to WRKY transcription factor 55
Os04g0177300	2108.4	13.6	1.40×10^{-10}	1.25×10^{-6}	HIP116, Rad5p N-terminal domain containing protein
Os04g0403500	180.6	0.008	2.44×10^{-5}	0.020228994	NAD(P)-binding domain containing protein
Os04g0503600	3609.6	179.2	3.33×10^{-5}	0.025674616	Similar to OSIGBa0112M24.5 protein
Os04g0510600	453.998	0	8.17×10^{-6}	0.009359965	Tetratricopeptide-like helical domain containing protein
Os04g0584201	168.564	0.224	4.06×10^{-6}	0.005649724	Hypothetical protein
Os05g0164800	116.004	0	1.80×10^{-5}	0.016782781	Similar to zinc transporter 6, chloroplast precursor
Os05g0203800	564.274	0	5.27×10^{-8}	1.68×10^{-4}	Transcription factor, floral organ development
Os05g0244700	1109	30.044	4.42×10^{-5}	0.031722495	Aminotransferase, class IV family protein
Os05g0594200	414.276	0.2	1.82×10^{-7}	4.78×10^{-4}	Similar to cation/proton exchanger 1a
Os06g0520600	8126.6	1677.2	1.71×10^{-5}	0.016639221	Similar to zinc finger CCCH type domain containing protein ZFN-like 1
Os06g0591200	180.6	0	4.55×10^{-5}	0.031797243	Conserved hypothetical protein
Os07g0170000	944.174	70.8	6.90×10^{-5}	0.042206892	Similar to Brn1-like protein
Os07g0668900	87.432	0	5.52×10^{-5}	0.036260256	Similar to serine/threonine-protein kinase PBS1
Os08g0191300	4671	468.6	6.61×10^{-9}	3.22×10^{-5}	Conserved hypothetical protein
Os08g0197300	160.8	0.4	2.59×10^{-5}	0.021001501	F-box domain, cyclin-like domain containing protein
Os08g0224700	3489.578	787.78	3.25×10^{-5}	0.02549321	Similar to 26S proteasome subunit RPN2a
Os08g0406900	989.088	0.6	1.39×10^{-9}	1.03×10^{-5}	Hypothetical protein
Os09g0133600	1001.524	0	3.80×10^{-20}	1.70×10^{-15}	Fibrillin, plastoglobule (PG) formation and lipid metabolism in chloroplasts
Os09g0433650	2942	197.4	3.16×10^{-5}	0.025214678	Tobacco mosaic virus coat protein family protein
Os09g0498700	241	1.6	1.78×10^{-5}	0.016782781	F-box domain, cyclin-like domain containing protein
Os09g0549300	214.2	0	4.47×10^{-5}	0.031722495	Flavin-containing monooxygenase FMO family protein
Os09g0552600	2250.304	264.8	1.20×10^{-7}	3.36×10^{-4}	RmlC-like jelly roll fold domain containing protein
Os10g0552400	1476.6	1.8	1.82×10^{-16}	4.07×10^{-12}	U-box E3 ubiquitin ligase
Os11g0213000	791.614	2.6	4.66×10^{-7}	9.91×10^{-4}	Similar to protein kinase domain containing protein, expressed
Os11g0295900	2590.4	134.4	3.26×10^{-6}	0.005025349	AP2-transcription factor, initiation of zygotic development
Os11g0437600	612.6	0.2	1.21×10^{-8}	4.17×10^{-5}	Protein of unknown function DUF506, plant family protein
Os11g0619800	220.4	0	2.60×10^{-7}	6.45×10^{-4}	Kelch related domain containing protein
Os12g0135800	316.184	0	9.14×10^{-9}	3.40×10^{-5}	Alpha/beta hydrolase fold-3 domain containing protein
Os12g0268000	585.318	1.658	1.17×10^{-6}	0.002272031	Cytochrome P450 monooxygenase, tryptamine 5-hydroxylase
Os12g0283400	1103.2	0.2	2.98×10^{-6}	0.004756237	Pectinesterase inhibitor domain containing protein
Os12g0637100	619.8	6.4	8.01×10^{-9}	3.25×10^{-5}	Similar to purple acid phosphatase

To analyze mis-expressed genes in polyspermic zygotes, gene expression profiles were compared between genes down-regulated in diploid zygotes after fertilization [29] and those up-regulated in polyspermic zygotes relative to diploid zygotes (Table 2). Two genes, Os01g0760000 and Os09g0551600, were identified as overlapped genes (Figure 5A), and, interestingly, Os09g0551600 encoded nucleasome/chromatin assembly factor D protein of HMG protein family. Next, comparison of gene expression profiles was conducted among genes up-regulated in diploid zygotes after fertilization [29], genes down-regulated in polyspermic zygotes relative to diploid zygotes (Table 3) and genes up-regulated in diploid zygotes after fertilization with paternal allele dependent expression [29]. Only one gene, Os11g0295900, was detected in diagram area overlapped with three gene groups (Figure 5B). Notably, the gene encoded *Oryza sativa Apospory-specific Genome Region (ASGR)- BABY-BOOM LIKE (BBML) 1 (OsASGR-BBML1)* (Table 3), which is a possible initiation factor that is important for zygotic development [29,30].

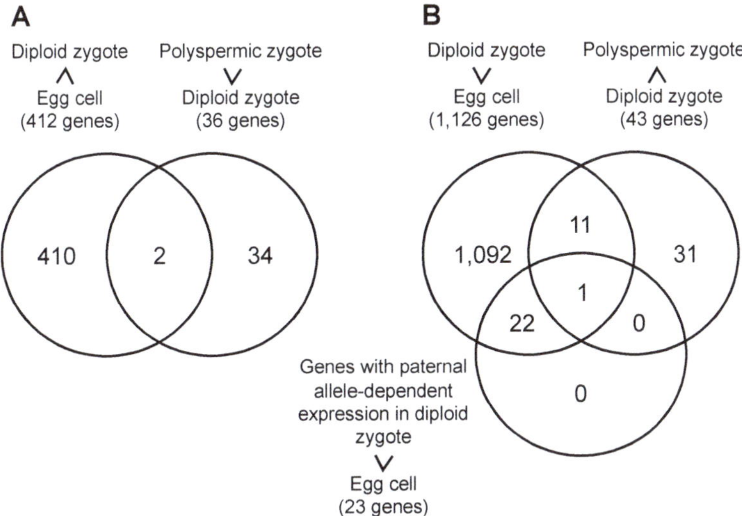

Figure 5. Gene expression in rice polyspermic zygotes and diploid zygotes. (**A**) Venn diagram of 412 genes, whose expression levels in diploid zygotes are suppressed after fertilization [29], and 36 genes, whose expressions are up-regulated in polyspermic zygotes relative to diploid zygotes (Table 2). (**B**) Venn diagram of 1,126 genes, which were detected as fertilization-induced genes in rice diploid zygotes [29], 43 genes, whose expressions are down-regulated in polyspermic zygotes relative to diploid zygotes (Table 3), and 23 genes which are up-regulated in diploid zygotes after fertilization with paternal allele dependent expression [29].

3. Discussion

Paternal genome excess appears to adversely affect polyspermic zygote development mainly during or after the completion of karyogamy. Interestingly, global *de novo* gene expression, termed zygotic genome activation (ZGA), is initiated in rice zygotes during or immediately after karyogamy is completed [31]. Thus, the developmental dysfunction of polyspermic zygotes was predicted to be due to the misexpression of genes important for zygotic development. In addition to gene expression profiles, chromatin/chromosome organization is also considered to be closely associated with plant cell developmental properties [32–34]. In a recent study involving chromatin conformation capture (3C) and high-throughput 3C (Hi-C) assays, Zhou et al. (2019) indicated that three-dimensional (3D) genomes of rice egg cells contain a compact silent center (CSC), and that the CSC appears to be reorganized after fertilization and the CSC reorganization may be involved in the regulation of ZGA [35]. The double dose of male chromatin in the nucleus of polyspermic zygotes may affect the 3D genome structure, resulting in abnormal ZGA. Notably, the genes that were more highly expressed in polyspermic rice zygotes than in diploid zygotes were enriched with molecular functions related to chromatin/chromosome assembly/organization (Table 2, Supplemental Table S2). In particular, expression level of a gene Os09g0551600, encoding nucleasome/chromatin assembly factor D protein of HMG protein family, appeared to be extremely high in polyspermic zygote compared to diploid zygotes, in which its expression level is suppressed after gamete fusion. The production of the molecular components required for chromatin/chromosome assembly/organization may be abnormally increased in polyspermic zygotes because of the 3D structure of the paternal excess genome content resulting from the double dose of the male genome. In addition, alternation of genome modification, including DNA methylation and histone acetylation/methylation, in polyspermic zygotes may be a reason for their developmental arrest, since epigenetic reprogramming is supposed to occur during development of zygotes [36,37].

In our previous study for investigating synergistic function of parental genomes in rice zygotes, 23 genes that were preferentially expressed from paternal allele were identified, and it was suggested that monoallelic or preferential gene expression from the paternal genome in the zygote is a safety mechanism for the egg cell, allowing it to suppress the gene expression cascade toward embryogenesis that is normally triggered by fusion with a sperm cell [29]. Therefore, we examined whether misexpression of these 23 genes in polyspermic zygotes occurs or not (Figure 5B). The results indicated that expression level of *OsASGR-BBML1* is highly suppressed in polyspermic zygotes relative to diploid zygotes. *OsASGR-BBML1*, which is alternatively named *OsBBML1*, encodes an AP2 transcription factor that is expressed in a paternal allele-dependent manner in rice zygotes to initiate zygotic development [29,30]. However, the *OsASGR-BBML1* expression level was substantially lower in polyspermic zygotes than in diploid zygotes. Interestingly, it has been reported that the suppression of the OsASGR-BBML1 function in rice zygotes via the ectopic expression of the OsASGR-BBML1-SRDX dominant repressor resulted in the developmental arrest of diploid zygotes at the one-cell stage [29]. The expression of *OsASGR-BBML1* at low levels may result in dysfunctional polyspermic zygotes after karyogamy. The BBM-related transcription factors, including pearl millet *ASGR-BBML1 (PsASGR-BBML1)* [38,39] and *Brassica napus BBM (BnBBM)* [40], reportedly function as determinant factors affecting diverse developmental events in embryonic tissues/cells (e.g., somatic embryogenesis and parthenogenesis). Therefore, elucidating the gene expression cascade triggered by OsASGR-BBML1 in rice zygotes is critical for characterizing the mechanism underlying global embryonic properties as well as zygotic development. Investigations aimed at identifying the genes regulated by OsASGR-BBML1 are currently in progress in our laboratories.

4. Materials and Methods

4.1. Plant Materials and Gamete Isolation

Oryza sativa L. cv. Nipponbare was grown in an environmental chamber (K30-7248; Koito Industries, Yokohama, Japan) at 26 °C with a 13-h light/11-h dark photoperiod. Transformed rice plants expressing the histone H2B-GFP fusion protein were generated as previously described [41]. Egg cells and sperm cells were isolated from rice flowers using a published procedure [42].

4.2. Production and Culture of Diploid Zygotes and Paternal Excess Polyspermic Zygotes

Zygotes were prepared from gametes isolated from wild-type rice plants or transgenic rice plants expressing H2B-GFP. To prepare diploid zygotes, an isolated egg cell and a sperm cell were electro-fused as described [27]. Polyspermic zygotes were produced via a serial fusion between two sperm cells and an egg cell, as previously described [28].

The produced zygotes were cultured on a Millicell-CM insert (Merck KGaA, Darmstadt, Germany) as described [27,43].

4.3. Microscopic Analysis

Gametes, zygotes, and embryo-like structures were examined using the BX-71 inverted microscope (Olympus, Tokyo, Japan). The fluorescence of H2B-GFP proteins in cells was observed using the BX-71 inverted fluorescence microscope (Olympus) with 460–490 nm excitation and 510–550 nm emission wavelengths (U-MWIBA2 mirror unit; Olympus). Digital images of the gametes, zygotes, and their cell masses were obtained using a cooled charge-coupled device camera (Penguin 600CL; Pixcera, Los Gatos, CA, USA) and the InStudio software (Pixcera).

4.4. cDNA Synthesis, Library Preparation, and mRNA Sequencing

Diploid and polyspermic zygotes cultured for 4–5 h after gamete fusion were washed four times by transferring the cells into fresh droplets of mannitol solution adjusted to 450 mOsmol kg^{-1} H_2O on coverslips. Each zygote was then transferred into the lysis buffer supplied in the SMART-Seq HT Kit (Takara Bio, Shiga, Japan), after which the lysates were stored at $-80\,°C$ until used. cDNA was synthesized and amplified from the cell lysates using the SMART-Seq HT Kit (Takara Bio) according to the manufacturer's instructions. The resulting amplified cDNA was purified using the Agencourt AMPure XP beads kit (Beckman Coulter, Brea, CA, USA). The quality and quantity of the purified cDNA were determined by the Qubit 3 Fluorometer with a Qubit dsDNA HS Assay Kit (Thermo Scientific, Waltham, MA, USA) and the Agilent 2100 BioAnalyzer with a High Sensitivity DNA chip (Agilent Technologies, Santa Clara, CA, USA). Sequencing libraries were prepared from the amplified cDNA using the Nextera XT DNA Library Prep Kit (Illumina, San Diego, CA, USA), after which they were purified with the Agencourt AMPure XP beads kit. After verifying the quality and quantity of the purified libraries with the Qubit 3 Fluorometer and the Agilent 2100 BioAnalyzer, the libraries were sequenced on the Illumina HiSeqX platform (Illumina) at Macrogen-Japan (Kyoto, Japan) to produce 150-bp paired-end reads.

4.5. Analyses of Transcriptome Data

The quality of the Illumina reads was evaluated using FastQC [44]. Regarding the preprocessing of the reads, adapter, poly-A, and low-quality sequences were removed using Cutadapt [45]. The remaining high-quality reads were mapped to the Nipponbare transcript sequences available in RAP-DB [46,47] using RSEM [48] and Bowtie2 [49]. On the basis of the mapping data, the reads mapped to each transcript (TPM) were counted, after which the read count was converted to transcripts per million using RSEM.

The DEGs between the diploid and polyspermic zygotes were identified using TCC [50] of the R software. The number of reads mapped to each transcript was compared between the zygotes and the false discovery rates (FDRs; q-values) were obtained. Genes with an FDR < 0.05 were extracted as DEGs.

4.6. Semi-Quantitative RT-PCR

The cDNAs of diploid and polyspermic zygotes at 4–5 h after fusion were synthesized as described above, and used as templates for PCR reaction. For PCR, 1 µL of the cDNA (200 pg/µL) was used as the template in a 50 µL PCR reaction with 0.3 µM of primers using KOD-FX DNA polymerase (Toyobo, Osaka, Japan) as follows: 30 or 35 cycles of 98 °C for 10 s, 55 °C for 30 s, and 68 °C for 1 min. Expression of the ubiquitin gene (Os02g0161900) was monitored as an internal control. Primer information is presented in Supplementary Table S5.

Supplementary Materials: The following are available online at https://www.mdpi.com/2223-7747/10/2/255/s1, Table S1: Developmental profiles of diploid and polyspermic rice zygotes, Table S2: Identified genes with putatively up-regulated expression levels in polyspermic zygotes, Table S3:

Identified genes with putatively down-regulated expression levels in polyspermic zygotes, Table S4: GO enrichment analysis of up-regulated genes in polyspermic zygotes, Table S5: Primers used for semi-quantitative RT-PCR.

Author Contributions: Conceptualization, R.D., E.T., and T.O.; methodology, R.D. and E.T.; data analyses, S.K. and K.Y.; investigation, R.D., E.T., and S.K.; writing, T.O., E.T., and S.K.; supervision, T.O.; funding acquisition, T.O. and K.Y. All authors have read and agreed to the published version of the manuscript.

Funding: This work was supported, in part, by MEXT KAKENHI (Grant-in-Aid for Scientific Research on Innovation Areas, Grant Nos. 19H04868 to TO and 19H04870 to KY) and JSPS KAKENHI (Grant-in-Aid for Challenging Exploratory Research, Grant No. 20K21317 to TO).

Informed Consent Statement: Informed consent was obtained from all subjects involved in the study.

Data Availability Statement: The transcriptome data were deposited in the DDBJ Sequence Read Archive [51] with the accession number DRA011171.

Acknowledgments: We thank T. Mochizuki (Tokyo Metropolitan University) for isolating the rice egg cells, and the RIKEN Bio Resource Center (Tsukuba, Japan) for providing cultured rice cells (Oc line).

Conflicts of Interest: The authors declare no conflict of interest.

References

1. Raghavan, V. Some reflections on double fertilization, from its discovery to the present. *New Phytol.* **2003**, *159*, 565–583. [CrossRef]
2. Nawaschin, S. Revision der Befruchtungsvorgange bei *Lilium martagon* und *Fritillaria tenella*. *Bull. Sci. Acad. Imp. Sci. Saint Pétersbourg* **1898**, *9*, 377–382.
3. Guignard, M.L. Sur les antherozoides et la double copulation sexuelle chez les vegetaux angiosperms. *Rev. Gén. Bot.* **1899**, *11*, 129–135.
4. Russell, S.D. Double fertilization. *Int. Rev. Cytol.* **1992**, *140*, 357–390.
5. Haig, D.; Westoby, M. Genomic imprinting in endosperm: Its effect on seed development in crosses between species, and between different ploidies of the same species, and its implications for the evolution of apomixis. *Philo. Trans. Biol. Sci.* **1991**, *333*, 1–13.
6. Scott, R.J.; Spielman, M.; Bailey, J.; Dickinson, H.G. Parent-of-origin effects on seed development in *Arabidopsis thaliana*. *Development* **1998**, *125*, 3329–3341.
7. Tiwari, S.; Spielman, M.; Schulz, R.; Oakey, R.J.; Kelsey, G.; Salazar, A.; Zhang, K.; Pennell, R.; Scott, R.J. Transcriptional profiles underlying parent-of-origin effects in seeds of *Arabidopsis thaliana*. *BMC Plant Biol.* **2010**, *10*, 72. [CrossRef]
8. Lu, J.; Zhang, C.; Baulcombe, D.C.; Chen, Z.J. Maternal siRNAs as regulators of parental genome imbalance and gene expression in endosperm of *Arabidopsis* seeds. *Proc. Natl. Acad. Sci. USA* **2012**, *109*, 5529–5534. [CrossRef]
9. Sekine, D.; Ohnishi, T.; Furuumi, H.; Ono, A.; Yamada, T.; Kurata, N.; Kinoshita, T. Dissection of two major components of the post-zygotic hybridization barrier in rice endosperm. *Plant J.* **2013**, *76*, 792–799. [CrossRef]
10. Toda, E.; Ohnishi, Y.; Okamoto, T. Effects of an imbalanced parental genome ratio on development of rice zygotes. *J. Exp. Bot.* **2018**, *69*, 2609–2619. [CrossRef]
11. Toda, E.; Okamoto, T. Polyspermy in angiosperms: Its contribution to polyploid formation and speciation. *Mol. Reprod. Dev.* **2020**, *87*, 374–379. [CrossRef] [PubMed]
12. Del Toro-De León, G.; García-Aguilar, M.; Gillmor, C.S. Non-equivalent contributions of maternal and paternal genomes to early plant embryogenesis. *Nature* **2014**, *514*, 624–627. [CrossRef] [PubMed]
13. Luo, A.; Shi, C.; Zhang, L.; Sun, M.-X. The expression and roles of parent-of-origin genes in early embryogenesis of angiosperms. *Front. Plant Sci.* **2014**, *5*, 729. [CrossRef] [PubMed]
14. Baroux, C.; Grossniklaus, U. The maternal-to-zygotic transition in flowering plants: Evidence, mechanisms, and plasticity. *Curr. Top. Dev. Biol.* **2015**, *113*, 351–371. [PubMed]
15. Anderson, S.N.; Johnson, C.S.; Chesnut, J.; Jones, D.S.; Khanday, I.; Woodhouse, M.; Li, C.; Conrad, L.J.; Russell, S.D.; Sundaresan, V. The zygotic transition is initiated in unicellular plant zygotes with asymmetric activation of parental genomes. *Dev. Cell* **2017**, *43*, 349–358. [CrossRef] [PubMed]
16. Ueda, M.; Aichinger, E.; Gong, W.; Groot, E.; Verstraeten, I.; Vu, L.D.; De Smet, I.; Higashiyama, T.; Umeda, M.; Laux, T. Transcriptional integration of paternal and maternal factors in the *Arabidopsis* zygote. *Genes Dev.* **2017**, *31*, 617–627. [CrossRef]
17. Zhao, P.; Begcy, K.; Dresselhaus, T.; Sun, M.-X. Does early embryogenesis in eudicots and monocots involve the same mechanism and molecular players? *Plant Physiol.* **2017**, *173*, 130–142. [CrossRef]
18. Zhao, P.; Zhou, X.; Shen, K.; Liu, Z.; Cheng, T.; Liu, D.; Cheng, Y.; Peng, X.; Sun, M.-X. Two-step maternal-to-zygotic transition with two-phase parental genome contributions. *Dev. Cell* **2019**, *49*, 882–893. [CrossRef]
19. Shen, W.H.; Parmentier, Y.; Hellmann, H.; Lechner, E.; Dong, A.; Masson, J.; Granier, F.; Lepiniec, L.; Estelle, M.; Genschik, P. Null mutation of *AtCUL1* causes arrest in early embryogenesis in *Arabidopsis*. *Mol. Biol. Cell* **2002**, *13*, 1916–1928. [CrossRef]

20. Xu, J.; Zhang, H.Y.; Xie, C.H.; Xue, H.W.; Dijkhuis, P.; Liu, C.M. EMBRYONIC FACTOR 1 encodes an AMP deaminase and is essential for the zygote to embryo transition in *Arabidopsis*. *Plant J.* **2005**, *42*, 743–756. [CrossRef]

21. Ronceret, A.; Gadea-Vacas, J.; Guilleminot, J.; Lincker, F.; Delorme, V.; Lahmy, S.; Pelletier, G.; Chabouté, M.E.; Devic, M. The first zygotic division in *Arabidopsis* requires de novo transcription of thymidylate kinase. *Plant J.* **2008**, *53*, 776–789. [CrossRef] [PubMed]

22. Andreuzza, S.; Li, J.; Guitton, A.E.; Faure, J.E.; Casanova, S.; Park, J.S.; Choi, Y.; Chen, Z.; Berger, F. DNA LIGASE I exerts a maternal effect on seed development in *Arabidopsis* thaliana. *Development* **2010**, *137*, 73–81. [CrossRef] [PubMed]

23. Yu, D.; Jiang, L.; Gong, H.; Liu, C.M. *EMBRYONIC FACTOR 19* encodes a pentatricopeptide repeat protein that is essential for the initiation of zygotic embryogenesis in *Arabidopsis*. *J. Integr. Plant Biol.* **2012**, *54*, 55–64. [CrossRef] [PubMed]

24. Guo, L.; Jiang, L.; Zhang, Y.; Lu, X.L.; Xie, Q.; Weijers, D.; Liu, C.M. The anaphase-promoting complex initiates zygote division in *Arabidopsis* through degradation of cyclin B1. *Plant J.* **2016**, *86*, 161–174. [CrossRef] [PubMed]

25. Yu, T.Y.; Shi, D.Q.; Jia, P.F.; Tang, J.; Li, H.J.; Liu, J.; Yang, W.C. The Arabidopsis receptor kinase ZAR1 is required for zygote asymmetric division and its daughter cell fate. *PLoS Genet.* **2016**, *12*, e1005933. [CrossRef]

26. Yang, K.J.; Guo, L.; Hou, X.L.; Gong, H.Q.; Liu, C.M. *ZYGOTE-ARREST 3* that encodes the tRNA ligase is essential for zygote division in *Arabidopsis*. *J. Integr. Plant Biol.* **2017**, *59*, 680–692. [CrossRef]

27. Uchiumi, T.; Uemura, I.; Okamoto, T. Establishment of an in vitro fertilization system in rice (*Oryza sativa* L.). *Planta* **2007**, *226*, 581–589. [CrossRef]

28. Toda, E.; Ohnishi, Y.; Okamoto, T. Development of polyspermic rice zygotes. *Plant Physiol.* **2016**, *171*, 206–214. [CrossRef]

29. Rahman, M.H.; Toda, E.; Kobayashi, M.; Kudo, T.; Koshimizu, S.; Takahara, M.; Iwami, M.; Watanabe, Y.; Sekimoto, H.; Yano, K.; et al. Expression of genes from paternal alleles in rice zygotes and involvement of OsASGR-BBML1 in initiation of zygotic development. *Plant Cell Physiol.* **2019**, *60*, 725–737. [CrossRef]

30. Khanday, I.; Skinner, D.; Yang, B.; Mercier, R.; Sundaresan, V. A male-expressed rice embryogenic trigger redirected for asexual propagation through seeds. *Nature* **2019**, *565*, 91–95. [CrossRef]

31. Ohnishi, Y.; Hoshino, R.; Okamoto, T. Dynamics of male and female chromatin during karyogamy in rice zygotes. *Plant Physiol.* **2014**, *165*, 1533–1543. [CrossRef] [PubMed]

32. Dong, P.; Tu, X.; Chu, P.Y.; Lü, P.; Zhu, N.; Grierson, D.; Du, B.; Li, P.; Zhong, S. 3D chromatin architecture of large plant genomes determined by local A/B compartments. *Mol. Plant* **2017**, *10*, 1497–1509. [CrossRef] [PubMed]

33. Liu, C.; Cheng, Y.J.; Wang, J.W.; Weigel, D. Prominent topologically associated domains differentiate global chromatin packing in rice from *Arabidopsis*. *Nat. Plants* **2017**, *3*, 742–748. [CrossRef] [PubMed]

34. Wang, M.; Wang, P.; Lin, M.; Ye, Z.; Li, G.; Tu, L.; Shen, C.; Li, J.; Yang, Q.; Zhang, X. Evolutionary dynamics of 3D genome architecture following polyploidization in cotton. *Nat. Plants* **2018**, *4*, 90–97. [CrossRef] [PubMed]

35. Zhou, S.; Jiang, W.; Zhao, Y.; Zhou, D.X. Single-cell three-dimensional genome structures of rice gametes and unicellular zygotes. *Nat. Plants* **2019**, *5*, 795–800. [CrossRef]

36. Kawashima, T.; Berger, F. Epigenetic reprogramming in plant sexual reproduction. *Nat. Rev. Gen.* **2014**, *15*, 613–624. [CrossRef]

37. Gehring, M. Epigenetic dynamics during flowering plant reproduction: Evidence for reprogramming? *New Phytol.* **2019**, *224*, 91–96. [CrossRef]

38. Conner, J.A.; Mookkan, M.; Huo, H.; Chae, K.; Ozias-Akins, P. A parthenogenesis gene of apomict origin elicits embryo formation from unfertilized eggs in a sexual plant. *Proc. Natl. Acad. Sci. USA* **2015**, *112*, 11205–11210. [CrossRef]

39. Conner, J.A.; Podio, M.; Ozias-Akins, P. Haploid embryo production in rice and maize induced by *PsASGR-BBML* transgenes. *Plant Reprod.* **2017**, *30*, 41–52. [CrossRef]

40. Boutilier, K.; Offringa, R.; Sharma, V.K.; Kieft, H.; Ouellet, T.; Zhang, L.; Hattori, J.; Liu, C.M.; Van Lammeren, A.A.; Miki, B.L.; et al. Ectopic expression of BABY BOOM triggers a conversion from vegetative to embryonic growth. *Plant Cell* **2002**, *14*, 1737–1749. [CrossRef]

41. Abiko, M.; Maeda, H.; Tamura, K.; Hara-Nishimura, I.; Okamoto, T. Gene expression profiles in rice gametes and zygotes: Identification of gamete-enriched genes and up- or down-regulated genes in zygotes after fertilization. *J. Exp. Bot.* **2013**, *64*, 1927–1940. [CrossRef] [PubMed]

42. Uchiumi, T.; Komatsu, S.; Koshiba, T.; Okamoto, T. Isolation of gametes and central cells from *Oryza sativa* L. *Sex. Plant Reprod.* **2006**, *19*, 37–45. [CrossRef]

43. Rahman, M.H.; Toda, E.; Okamoto, T. In vitro production of zygotes by electro-fusion of rice gametes. *Methods Mol. Biol.* **2020**, *2122*, 257–267. [PubMed]

44. Simon, A. FastQC: A Quality Control Tool for High Throughput Sequence Data. 2010. Available online: https://www.bioinformatics.babraham.ac.uk/projects (accessed on 17 December 2019).

45. Martin, M. Cutadapt removes adapter sequences from high-throughput sequencing reads. *EMBnet J.* **2011**, *17*, 10–12. [CrossRef]

46. Sakai, H.; Lee, S.S.; Tanaka, T.; Numa, H.; Kim, J.; Kawahara, Y.; Wakimoto, H.; Yang, C.C.; Iwamoto, M.; Abe, T.; et al. Rice Annotation Project Database (RAP-DB): An integrative and interactive database for rice genomics. *Plant Cell Physiol.* **2013**, *54*, e6. [CrossRef]

47. Kawahara, Y.; De la Bastide, M.; Hamilton, J.P.; Kanamori, H.; McCombie, W.R.; Ouyang, S.; Schwartz, D.C.; Tanaka, T.; Wu, J.; Zhou, S.; et al. Improvement of the *Oryza sativa* Nipponbare reference genome using next generation sequence and optical map data. *Rice* **2013**, *6*, 4. [CrossRef]

48. Li, B.; Dewey, C.N. RSEM: Accurate transcript quantification from RNA-Seq data with or without a reference genome. *BMC Bioinform.* **2011**, *12*, 323. [CrossRef]
49. Langmead, B.; Salzberg, S.L. Fast gapped-read alignment with Bowtie 2. *Nat. Methods* **2012**, *9*, 357–359. [CrossRef]
50. Sun, J.; Nishiyama, T.; Shimizu, K.; Kadota, K. TCC: An R package for comparing tag count data with robust normalization strategies. *BMC Bioinform.* **2013**, *14*, 219. [CrossRef]
51. Kodama, Y.; Shumway, M.; Leinonen, R. The sequence read archive: Explosive growth of sequencing data. *Nucleic Acids Res.* **2012**, *40*, D54–D56. [CrossRef]

Article

Indirect Somatic Embryogenesis and Cryopreservation of *Agave tequilana* Weber Cultivar 'Chato'

Lourdes Delgado-Aceves [1], María Teresa González-Arnao [2], Fernando Santacruz-Ruvalcaba [1], Raquel Folgado [3] and Liberato Portillo [1,*]

[1] Centro Universitario de Ciencias Biológicas y Agropecuarias, Universidad de Guadalajara, Zapopan 45200, Mexico; bmlda108@gmail.com (L.D.-A.); fernando.santacruz@academicos.udg.mx (F.S.-R.)

[2] Laboratorio de Biotecnología y Criobiología Vegetal, Facultad de Ciencias Químicas, Universidad Veracruzana, Orizaba 94340, Mexico; teregonzalez@uv.com.mx

[3] Huntington Library, Art Museum, and Botanical Gardens, San Marino, CA 91108, USA; rfolgado@huntington.org

* Correspondence: portillo@cencar.udg.mx; Tel.: +52-333-777-1192

Abstract: *Agave tequilana* Weber cultivar 'Chato' represents an important genetic supply of wild severely in decline populations of 'Chato' for breeding and transformation programs. In this work, the indirect somatic embryogenesis and cryopreservation of Somatic Embryos (SEs) were investigated using the 'Chato' cultivar as a study case. Methods: Embryogenic calli were induced by the cultivation of 1 cm of young leaves from in vitro plants on MS semisolid medium supplemented with 24.84, 33.13, 41.41, 49.69, and 57.98 µM 4-amino-3,5,6-trichloro-2- pyridinecarboxylic acid (picloram) in combination with 2.21, 3.32, and 4.43 µM 6-benzylaminopurine (BAP). The origin and structure of formed SEs were verified by histological analysis. Cryopreservation studies of SEs were performed following the V-cryoplate technique and using for dehydration two vitrification solutions (PVS2 and PVS3). Results: The highest average (52.43 ± 5.74) of produced SEs and the Embryo Forming Capacity (estimated index 52.43) were obtained using 49.69 µM picloram and 3.32 µM BAP in the culture medium. The highest post-cryopreservation regrowth (83%) and plant conversion rate (around 70%) were achieved with PVS2 at 0 °C for 15 min. Conclusion: Our work provides new advances about somatic embryogenesis in *Agave* and reports the first results on cryopreservation of SEs of this species.

Keywords: regeneration; picloram; cryoplate; vitrification solutions; long-term preservation

Citation: Delgado-Aceves, L.; González-Arnao, M.T.; Santacruz-Ruvalcaba, F.; Folgado, R.; Portillo, L. Indirect Somatic Embryogenesis and Cryopreservation of *Agave tequilana* Weber Cultivar 'Chato'. *Plants* **2021**, *10*, 249. https://doi.org/10.3390/plants10020249

Academic Editors: Minako Ueda and Sharyn Perry

Received: 5 December 2020

Accepted: 21 January 2021

Published: 28 January 2021

Publisher's Note: MDPI stays neutral with regard to jurisdictional claims in published maps and institutional affiliations.

1. Introduction

Agave plants are distributed in several wild and cultivated areas of different Mexican states, as well as preserved by many local human populations [1]. There are various cultivars of *A. tequilana* that were previously used to make tequila, among which are 'Azul', 'Chato', 'Chino', 'Pata de mula', 'Mano larga', 'Bermejo', 'Xigüin', and 'Moraleño' [2,3]. Among them, *A. tequilana* 'Chato' is a valuable resource, of which wild populations are severely displaced due to overexploitation of specific cultivar (*A. tequilana* 'Azul' monocrops). The growing demand for products derived from *Agave* spp. increases the need for plantations; however, most species of economic interest are severely affected at the risk of loss by the intense pressure and use of these plant natural resources. Besides, the slow growth to reach its sexual reproductive stage (8–15 years) has made this resource particularly vulnerable since the plants are usually exploited before the formation of the flower stalk, avoiding the dissemination of seeds and reducing variability [4,5]. *Agave* plant propagation is primarily achieved by the multiplication of rhizomatous shoots, which arise from the basal stem of parent plants. *A. tequilana* 'Chato' is considered a key cultivar to provide variability to species of commercial use and contributes to the diversify of their populations [6]. However, *A. tequilana* 'Azul' is the only approved cultivar by the Mexican regulation to produce tequila [7], and therefore, the only one with the protection of origin denomination [6,8]. This implies that the tequila industry ignores the genetic resources of other important local cultivars.

The application of tissue culture techniques in *Agave* spp. has been useful to promote large-scale plant production of endangered and economically important species [9–17]. Somatic embryogenesis represents a valuable in vitro regeneration system [14] to produce somatic embryos (SEs), which are essential for breeding and genetic transformation programs, as well as for germplasm conservation. In addition, the histological studies represent an important complementary approach to follow up and verify the course of the embryogenic process from the cellular origin of SE until germination of the seedling [18]. The developmental stages of the unicellular origin of somatic embryogenesis for *A. tequilana* have been previously presented [14].

SEs as biotechnological products can be long-term preserved only by using cryopreservation techniques [19]. The cryogenic storage (i.e., preservation at ultralow temperature, mainly in liquid nitrogen (LN), $-196\,^\circ$C) of complex structures with a heterogeneous cellular composition like somatic embryos has been mostly achieved using different vitrification-based procedures [20,21]. A common characteristic of these techniques is that dehydration prior to cooling is the critical step to induce vitrification, which means the transition of the aqueous contents of tissues directly to the amorphous glass phase during the rapid or ultrarapid immersion in LN. The development of the V-cryoplate method [22], which involves the encapsulation of plant material over aluminum cryoplates, allowing the manipulation of many samples at the same time along the different stages of the protocol. In addition, it ensures ultrarapid rates of cooling and warming, which help to improve the post-cryopreservation recovery and, consequently, the effectiveness of the procedure [22,23]. Following the V-cryoplate approach, samples are osmotically dehydrated by exposure to Plant Vitrification Solutions (PVS), which are highly concentrated mixes of cryoprotectants. PVS2 formulation [24] has proved to be the most effective for dehydration of the tissue of different plant species, while PVS3 has been demonstrated to be less toxic and very convenient when PVS2 results cytotoxic [25].

The integration of somatic embryogenesis and cryopreservation has been successfully achieved in some species, such as olive [26], cocoa [27], and avocado [28]. However, so far, there are no reports of its application to *Agave* embryogenic cultures.

This work aimed to induce the somatic embryogenesis in *A. tequilana* 'Chato', validate the embryogenic process by histological analysis, and study the cryopreservation of agave SEs following the V-cryoplate procedure.

2. Results

2.1. Indirect Somatic Embryogenesis

Produced calli formed clumps and were creamy in color. The embryogenic calli were friable and contained numerous elongated, spherical units that formed translucent and immature SEs. The results obtained under the factors analysis that included the auxin picloram and the cytokinin 6-benzylaminopurine (BAP) showed a highly significant difference as well in their interaction ($p < 0.01$). Since an interaction was detected between factors, all growth regulator combinations were analyzed and were found to have a significant effect ($p < 0.01$) on the differentiation response of calli to produce SEs. Table 1 shows the least significant difference (LSD) test for the mean number of somatic embryos ($p < 0.05$). The significantly highest number of SEs (52.43 ± 5.74 SEs) was achieved by adding 49.69 µM picloram and 3.32 µM BAP to the induction culture medium. The formation of SEs significantly decreased using other combinations. The same concentration (49.69 µM) of picloram gave the best result (24 SEs) when used with the lowest (2.21 µM) concentration of BAP. By contrast, the highest concentration of BAP only had a minor effect (16 SEs) at a lower picloram concentration (33.13 µM). The Embryo Forming Capacity (EFC) of callus from 1 cm^2 of young leaves derived from in vitro plants of 'Chato' cv. ranged from 1.29 up to a maximum value of 52 after 60 days of culturing at the best combination of picloram (49.69 µM) with BAP (3.32 µM).

Table 1. Effect of plant growth regulators on the differentiation rate of somatic embryos derived from calli of in vitro young leaves of *Agave tequilana* 'Chato'.

Treatment	Concentration of Growth Regulators (µM)		Calli that Formed SEs %	Number of Somatic Embryos (Mean ± se) *		EFC
	Picloram	BAP				
1	24.84	2.21	56.25	10.93 ± 4.58	fg	6.15
2	33.13	2.21	68.75	16.18 ± 6.30	def	11.12
3	41.41	2.21	31.25	10.81 ± 9.17	fg	3.37
4	49.69	2.21	56.25	24.12 ± 4.00	bc	13.57
5	57.98	2.21	43.75	20.25 ± 5.61	cde	8.85
6	24.84	3.32	31.25	5.06 ± 2.29	g	1.58
7	33.13	3.32	62.50	13.37 ± 5.34	ef	8.35
8	41.41	3.32	93.75	29.43 ± 5.30	b	27.59
9	49.69	3.32	100.00	52.43 ± 5.74	a	52.43
10	57.98	3.32	75.00	21.87 ± 6.19	bcd	16.40
11	24.84	4.43	56.25	27.18 ± 2.33	bc	15.29
12	33.13	4.43	43.75	16.00 ± 4.15	def	7.00
13	41.41	4.43	31.25	4.56 ± 2.27	g	1.42
14	49.69	4.43	25.00	5.18 ± 1.40	g	1.29
15	57.98	4.43	37.50	9.37 ± 8.92	fg	3.51

* Data represent the means ± standard error (se) of number of somatic embryos (SEs) per explant of 1 cm^2. Data within columns with the same letter are not significantly different, i.e., $p < 0.05$ (least significant difference (LSD) test). EFC: Embryo Forming Capacity.

The first step of the embryogenic process was the cellular disorganization of the leaf (Figure 1a,b), followed by the formation of abundant calli (Figure 1c), and the asynchronous proliferation of SEs on the embryogenic callus surface (Figure 1d). The typical structures of the developmental phases of SE were observed: globular (Figure 1e,f), scutellar (Figure 1g,h), and coleoptilar (Figure 1i,j). The maturation of SEs from the globular stage to germination and root development (Figure 1e–p) showed a transition time of approximately eight weeks. The conversion rate of SEs to ex vitro (acclimated) plants (Figure 1q) was 92%.

2.2. Histological Observations

The staining used in the present study characterized embryogenic calli by discriminating small dividing cells with a higher proportion of nucleic acids (stained with acetocarmine due to their affinity to this pigment) from those large and vacuolated (stained with Evan's blue due to the basic pH). The unicellular origin of the SEs is indicated by an initial asymmetrical cell division (Figure 2a). Pre-embryogenic cells showed properties that are common to cells in the division stage with high metabolic activity (Figure 2b). A suspensor type structure independent of the embryogenic callus could be observed in the next stage (Figure 2c,d). At this stage, we also observed a potential hypophysis like structure, which is a prominent cell zone (cells in contact with the embryo that link it to the suspensor) that promotes the formation of the radicle (Figure 2c). At the globular stage, the delimiting protoderm between the callus and the somatic embryo marked the independence of the embryo (Figure 2e). Then, differentiated pro-vascular strands were observed in scutellar embryos (Figure 2f). At the scutellar stage (Figure 2g,h), the apical axis or plumule (shoot) and basal axis that gives rise to radicle (root) differentiation (Figure 2j). The coleoptilar stage (Figure 2i,j) is characterized by the histodifferentiation of coleoptiles as the last major morphogenic transition of the embryos.

2.3. Cryopreservation by the V-cryoplate Method

Somatic embryogenesis is an asynchronous process that makes difficult the selection of a specific embryo size and of a determined physiological stage at a given culture time. After 60 days time on expression medium, we found that the most frequent sizes of SEs, ranged from 0.1 to 4.0 mm, and that globular and coleoptilar were the developmental stages mostly observed. Therefore, based on the higher frequency of these two factors, we selected SEs of 1.0 to 2.0 mm size at the coleoptilar stage to standardize the material choice for cryopreservation. SEs at earlier stages were subcultured on solid MS medium to stimulate their growth until obtaining material according to the pre-established parameters for cryogenic experiments.

Figure 1. Stages of indirect somatic embryogenesis in *Agave tequilana* Weber cultivar 'Chato'. Callus production in leaf after (**a**) 10 days, (**b**) 25 days, and (**c**) 40 days of culturing (bars 2.0 mm), (**d**), embryogenic callus in expression medium with somatic embryos (SEs) at several stages after 60 days of culturing (bar 2.0 mm), (**e**) and, (**f**), globular SEs without and with chlorophyll presence (bars 0.2 mm and 0.5 mm), respectively, (**g**) and (**h**) scutellar stage (bars 0.5 and 1.0 mm), respectively, (**i–o**), subsequent development of the colleoptilar stage presenting radicle origin. (**i,j**) (bars 1.0 mm and (**k–o**) bars 2.0 mm), respectively, (**p**) Somatic embryo elongation and tissue maturation (bar 5.0 mm), and (**q**) ex vitro-grown plant under greenhouse conditions (250 d) (bar 1.0 cm).

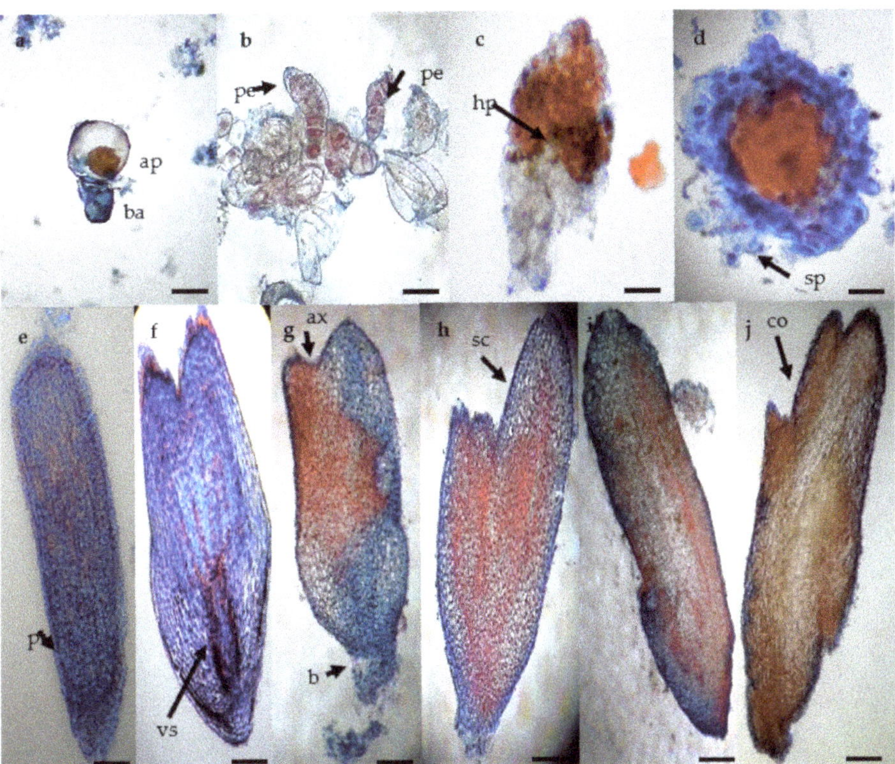

Figure 2. Histological analysis of somatic embryogenesis process in *Agave tequilana* Weber 'Chato'. (**a**) asymmetric cell division giving rise to small apical cell (red) and basal cell (blue) (bar 50 μm), (**b**) proembryos showing group of cells with one side of rapid cellular division (black arrow) (bar 100 μm), (**c**) late proembryo stage with dyed suspensor in blue and hypophysis (bar 100 μm), (**d**) specific formation pattern of the globular stage with vestigial suspensor structures (bar 100 μm), (**e**) late globular stage with the formation of the protoderm (bar 200 μm), (**f**) late globular stage (bar 200 μm), (**g**) early scutellar stage showing an apical axis (black arrow) (bar 200 μm), (**h**) late scutellar stage (bar 200 μm), (**i**) early coleoptillar stage (bar 200 μm), (**j**) late colleoptillar stage (bar 200 μm). Apical cell (ap), basal cell (ba), proembryo (pe), hypophysis (hp), suspensor (sp), protoderm (pt), pro-vascular strands (vs), apical axis (ax), basal axis (bx), scutellum (sc), and coleoptile (co).

Results of cryopreservation experiments expressed by the regrowth and plant conversion of SEs before and after immersion in liquid nitrogen (LN) are shown in Figures 3 and 4. Before cryopreservation, the regrowth of SEs was significantly ($p \leq 0.05$) influenced by the duration of exposure to both PVSs (PVS2 and PVS3) (Figure 3). Regrowth was significantly ($p \leq 0.05$) reduced (from 96% to 73%) after 30 min of exposure to PVS2 or 45 min exposure to PVS3 (reduction from 100% to 57%). However, conversion to plants was significantly affected after 15 min of treatment with either of the two PVSs used (Figure 4). After cryopreservation (+LN), regrowth and plant conversion were only detected in SEs that had been dehydrated with either PVS2 or PVS3. In general, SEs of *A. tequilana* 'Chato' tolerated all the dehydration durations assessed at low temperature (0 °C) with both PVS. The highest and significantly different percentages of regrowth and conversion to the plant were achieved after 15 min (83% and 73%, respectively) or 30 min (77% and 67%, respectively) of exposure to PVS2 and after 30 min (80% and 70%, respectively) to PVS3. Therefore, to achieve the best post-cryopreservation recovery expressed by the regrowth and plant conversion rates, a longer exposure time to PVS3 was required in comparison to that needed when PVS2 was used. Nevertheless, SEs treated with PVS3 showed faster regrowth during the first 30 days of reculture than those SEs treated with less or for the same time with PVS2. After 60 days of reculture, a similar appearance in plants cryopreserved was reached by all recovered SEs regardless of the PVS applied for dehydration.

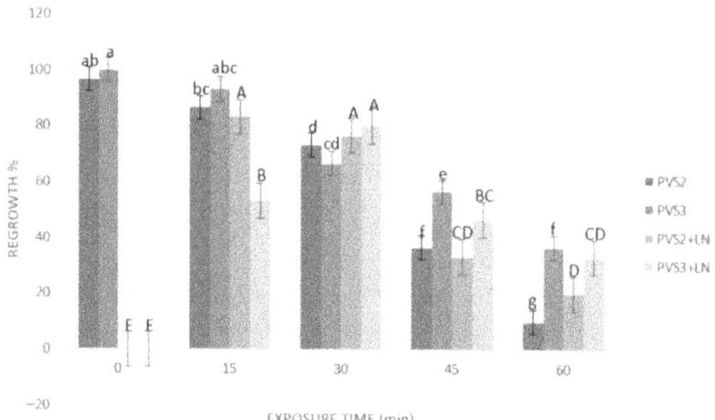

Figure 3. Effect of exposure length to plant vitrification solutions (PVS) PVS2 or PVS3 at 0 °C on regrowth of somatic embryos (SEs) before and after cryopreservation. SEs were precultured on MS solid medium with 0.3 M sucrose in the dark for one day, encapsulated over the cryoplate with calcium alginate (2%) containing 0.4 M sucrose, and loaded with 1 M sucrose + 2 M glycerol for 15 min before exposure to PVS. Data are represented by means (%) ± standard error (se). Bars with the same letter are not significantly different $p \leq 0.05$ (LSD test). Lowercase letters refer to non-cryopreserved controls, and uppercase letters refer to cryopreserved plus liquid nitrogen (+LN) samples. Regrowth was detected as the percentage of SEs that showed elongation of coleoptile and the formation of radicle 45 days after their transfer to the culture medium.

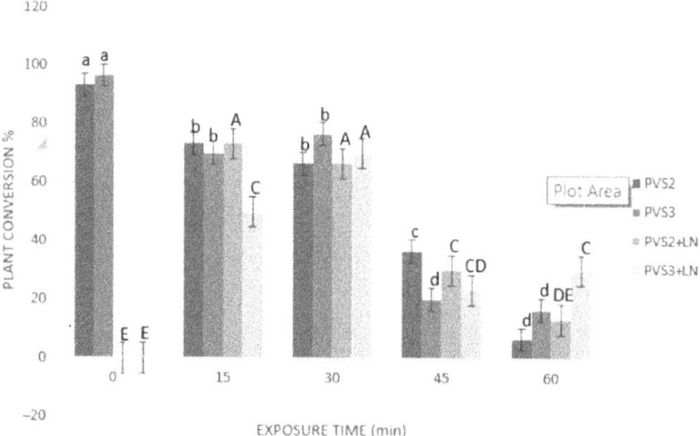

Figure 4. Effect of exposure length to PVS2 or PVS3 at 0 °C on the conversion of somatic embryos (SEs) to plants before and after cryopreservation. SEs were precultured on MS solid medium with 0.3 M sucrose in the dark for one day, encapsulated over the cryoplate with calcium alginate (2%) containing 0.4 M sucrose, and loaded with 1 M sucrose + 2 M glycerol for 15 min before exposure to PVS solution. Data are represented by means (%) ± standard error (se). Bars with the same letter are not significantly different $p \leq 0.05$ (LSD test). Lowercase letters refer to non-cryopreserved controls, and uppercase letters refer to cryopreserved (+LN) samples. Conversion to plants of cryopreserved SEs was evaluated after 105 days of culturing on MS medium.

Plants formed of cryopreserved SEs displayed normal growth and development (Figure 5), i.e., they were morphologically similar to control plants not subjected to cryopreservation and originating by germination of zygotic (Figure 5a) or somatic (Figure 5b,c) embryos, respectively.

Figure 5. Conversion to plants of *Agave tequilana* Weber 'Chato'. (**a**) Plant derived from zygotic embryo after 60 days of culture on MS medium (bar 2 cm), (**b**) Plant derived from non-cryopreserved SE after 60 days of culture on MS medium (bar 1 cm), (**c**) Plant derived from SE recovered after cryopreservation and cultivated for 105 days on MS medium (bar 1 cm).

3. Discussion

3.1. Indirect Somatic Embryogenesis

There are several factors involved in the acquisition of the embryogenic competence in explants cultivated in vitro, such as the balance of hormones, osmotic conditions, change of pH, concentrations of amino acids and salts, and treatments with various chemical substances [29–31]. The efficiency of induction depends not only on the culture conditions but also on the genotype, explant source, and its stage of development [32]. In *Agave* spp., somatic embryogenesis has been induced using different types of explants: roots [33], leaf [14], and putative basal part of the stem [34,35]. Regarding growth regulators, 2,4-D (2,4-diclorophenoxiacetic acid) was the most employed auxin in combination with other hormones to induce this process in several species: *A. victoria-reginae* [36], *A. sisalana* [13], *A. tequilana* 'Azul' [14], *A. vera-cruz* [37], *A. salmiana* [38], and *A. angustifolia* [17,39].

In our experiment, we replaced the use of 2,4-D with picloram, which in combination with BAP, allowed the indirect development of SEs in *A. tequilana* Weber cultivar 'Chato' at a similar rate as previously reported with 2,4-D [14]. Following this approach and under the best-determined culture conditions, the embryogenic calli reached an EFC of 52 with a high frequency of plant conversion (90%). On the other hand, Santiz et al. [40] reported a higher EFC index in *A. grivalgensis* when the concentration of cytokinin was higher than that of the auxin by combining BAP and 2,4-D. In contrast, the balance of auxin-cytokinin concentrations assayed in our study improved the callus embryogenic capacity when BAP was used at a lower concentration than picloram auxin (Table 1). The beneficial effect of picloram to induce somatic embryogenesis has also been demonstrated in the *A. americana* species [35], as well as in other plant species like *Urochloa* [41], *Gasteria verrucosa*, and *Haworthia fasciata* [42]. However, to our knowledge, picloram has not been used before to induce somatic embryogenesis from agave leaf explants.

The morphological evaluation performed during the different stages of development of the whole somatic embryogenesis process (Figure 1) demonstrated accordance with the one reported by Portillo et al. [14], who also used young leaf explants to generate SEs. In addition, it had similarities with the stages observed during the morphological development of zygotic embryos (ZE) of this species (Figure 1p,q). Furthermore, it was visually defined that germinated SEs produced normal plantlets, which resembled the ones obtained by germination of ZE. These observations are contrary to that obtained by Monja-Mio and Robert [34] in *A. fourcroydes* Lem., who reported having achieved direct somatic embryogenesis using the same hormone combination with other explant type; however, the morphology of their developing somatic embryos did not resemble the zygotic ones.

Regeneration systems for *Agave* spp. have not yet been characterized in detail. Somatic embryogenesis and organogenesis have been confused as the same processes [13,34–37]. Therefore, we emphasize the importance of ontogenetic observations by performing a comparison between SE and ZE of the studied species.

3.2. Histological Analysis

In this study, we have documented the unicellular origin and cellular structures (protoderm, scutellum, and coleoptile) in agave embryogenic development, which suggests crucial evidence for the regeneration via somatic embryogenesis versus organogenesis. Embryogenic cells from which embryoids are visibly derived shown a series of common characteristics as a high nucleus-cytoplasm ratio, thick cell wall, and small size (Figure 2a) resembling rapidly dividing meristematic cells (Figure 2b,c). According to Williams and Maheswaran [43], there are several points for discussion among the known regeneration systems, including the unicellular or multicellular origin, the physical or physiological independence of the starting embryogenic cells from the tissue of origin, the similarity or dissimilarity with zygotic embryogenesis, and the induction process controlled by exogenous growth regulators as opposed to the internal physiological state of the tissue explant.

Auxin concentrations higher than cytokinins (i.e., 9.0 or 13.6 µM 2,4-D, and 4.0 µM BAP) have resulted in somatic embryogenesis in *A. tequilana* [14]. It was also observed that a concentration of auxin higher than cytokinin correlates with the unicellular origin (Figure 2a), and the independence of SEs from the parent callus (Figure 2d,e), and its similar development to ZE (Figure 2j), as reported by Portillo et al. [14] and Ayala-González et al. [44]. It has been determined that regeneration can be achieved from meristemoids (organogenesis) when new shoots are induced from callus or directly upon explant tissues, or via SEs that resemble seed embryos developing to seedlings in the same way [43,45]. At this point, the question of the origin of one or several cells for SE is directly related to the coordinated behavior of neighboring cells as a morphogenetic group [43,45]. According to several authors, a somatic embryo is defined as a new individual that arises from a single cell and has no vascular connection to the parent cells; multicellular origin seems to produce embryoids fused with parent cells over a wide area of the root pole of the axis region [13,46], while a unicellular origin is more likely to produce individual embryoids with a narrower structure similar to a suspensor [30,43,45–47]. Therefore, the regeneration processes must be thoroughly studied in order to clarify and define whether a multicellular regenerate is a SE or an organ primordium. Halperin, emphasized-the use of histological sections in regeneration systems to determine the cellular origin, morphology, and stages of embryonic development [48]. By means of a histology study and morphology comparison, we found that in our system *Agave* SEs had a unicellular origin in agreement with previous work [14] (Figure 2a), and their development acquired in later stages were congruent (Figure 5a,b) [49].

3.3. Cryopreservation by V-cryoplate Method

Dehydration with PVS proved to be a critical step for cryopreservation of the agave SEs. The comparison of the two PVS (PVS2 and PVS3) showed that at the best exposure time, regrowth (about 80%) and plant conversion rate (about 70%) remained similar before and after cryopreservation, indicating that the immersion in LN caused no additional detrimental effects.

The positive effect of using a low temperature (0 °C) for dehydration with a PVS was first reported by Yamada et al. [50]. This approach of osmoprotection has been very useful with other tropical species, which usually are relatively sensitive to such dehydration treatments [51]. At this low temperature, PVS2 was demonstrated to be more effective than PVS3 in a shorter dehydration period. It was convenient to reduce the harmful effects due to the overexposure of tissues to PVS2. This effectiveness seems to be related to its chemical composition and lower viscosity in comparison with PVS3, which allows removal of the freezable water from cells [52] and increase the ability to vitrify during rapid immersion in LN [53].

Some studies on cryopreservation of SEs of avocado [54] and olive [55] have reported 30 min of dehydration as the optimal exposure time to PVS2 using a droplet-vitrification procedure. There are other cases in which more prolonged exposures (60–90 min) have proved to be beneficial [26,28,53]. However, our results showed that in the case of the SEs of cultivar 'Chato', increasing the exposure time beyond 30 min was detrimental, whichever PVS was used, provoking a significant drop in the regeneration response.

In this work, we determined suitable conditions to cryopreserve SEs of *Agave tequilana* cultivar 'Chato' following the V-cryoplate procedure and using two PVS (PVS2 or PVS3) up to 30 min at 0 °C. So far, the use of the V-cryoplate technique in SE has only been reported by Pettinelli et al. [56] to cryopreserve SEs derived from in vitro roots of guinea (*Petiveria alliacea*) and dehydrated with PVS2 for 15 min. These results match with the duration of osmoprotective treatment, which allowed them to obtain the best response after cryopreservation of agave SEs. Future adaptation of this protocol to other agave cultivars will depend on the water content and the sensitivity of their SEs to the PVS.

4. Materials and Methods

4.1. Induction of Indirect Somatic Embryogenesis

Rhizomatous shoots of *Agave tequilana* Weber cultivar 'Chato' were obtained from parent plants at physiological maturity (seven years old), which were provided by The Botanical Garden from the University of Guadalajara for initial in vitro shoot cultures [14]. Micropropagation of rhizomatous shoots was carried out in MS medium [57] supplemented with L2 vitamins [58] and 22.15 µM of 6-benzylaminopurine (BAP). The genotype 7 (SEC7) was selected due to its high frequency of proliferation of vigor shoots. The somatic embryogenesis process was followed and evaluated, defining three stages: callus induction, callus differentiation to SE, and conversion of obtained SEs to plants. For callus induction, segments (1 cm^2) adjusted with a millimetric ruler of young leaves from in vitro plants were cultivated for 40 days in a glass jar of 100 mL capacity with 25 mL of culture medium. The induction culture media comprised the MS basal formulation supplemented with 24.84, 33.13, 41.41, 49.69, or 57.98 µM of 4-amino-3,5,6-trichloro-2-pyridinecarboxylic acid (picloram) in combination with 2.21, 3.32, or 4.43 µM BAP (a 5 × 3 bifactorial design), 3% sucrose (27360 Golden Bell^{MR}), and solidified with 8 g L^{-1} agar (A-1296 Sigma$^®$). Four leaf segments were placed per culture jar (experimental unit) and for each combination of growth regulators. The experiment was replicated four times.

To induce callus differentiation (second stage), the masses of calli produced were transferred to Petri dishes with 25 mL of expression medium and cultured for 60 days. Expression medium consisted of a modified MS basal formulation [59] supplemented with

500 mg L^{-1} glutamine, 250 mg L^{-1} casein hydrolysate, 3% sucrose, and solidified with 6 g L^{-1} phytagel (P-8169 Sigma®) [14].

The pH of all culture media was adjusted to 5.8 ± 0.05, and then, they were sterilized in an autoclave at 121 °C with 1.3 kg cm^{-2} of pressure for 15 min. Cultures conditions for callus induction and differentiation into somatic embryos comprised the exposure of samples to a photoperiod (16/8 h light/dark), with a luminous intensity of 27 μmol m^{-2} s^{-1}, and the regulation of temperature at 25 ± 2 °C.

The number of generated SEs per callus was recorded after 60 days of culturing using a stereoscope with 10× magnification (Leica® microsystems EZ4 W). For all treatments, the means of percentages of embryogenic calli (which formed SEs) and from the number of formed SEs were estimated considering four explants per Petri dish. An index of *Embryo Forming Capacity* (EFC) was defined by adapting and modifying the equation previously reported [60].

EFC = (Average percentage of calli forming SE) (Average number of formed SE)/100

The final stage of the somatic embryogenesis process was performed by transferring one hundred SEs from the expression medium to MS medium without growth regulators for the other 60 days. For acclimation, rooted plantlets were then removed from the culture medium and placed in trays with a wet soil mixture of 7:3 (v/v) peat moss and perlite under greenhouse conditions with full sun at 27 ± 5 °C and 75% RH. The conversion rate of SEs to plants was calculated after the 60 days of culture on MS medium when the material was ready to be ex vitro transferred to be acclimated.

4.2. Histological Analysis

Histological studies were carried out to support the theories on the unicellular and multicellular origin, development, and characteristics of SE according to [61,62]. Fresh embryogenic calli (0.02 g) and SEs of the genotype SEC7 at different developmental stages were fixed using 70% v/v alcohol and embedded in polyethylene glycol (PEG, 1450 molecular mass) in a 1:4 proportion (PEG: deionized water) according to the protocol described [63]. The experiments were replicated six times. A rotatory microtome was used to obtain 15 μm sections from the samples in PEG; then, they were stained with a double treatment using acetocarmine 0.5% (1:1 w/v) and 0.5% Evan's blue (1:1 w/v) [64]. A light microscope was used to analyze the tissues.

4.3. Cryopreservation of Somatic Embryos

Experiments were performed following the V-cryoplate method [65] and using cryoplates with ten wells with oval shapes according to design No. 3 (37 mm length × 0.5 mm thickness, wells with 2.5 mm length, 1.5 mm width, and 0.75 mm depth), manufactured by the Japanese Company (Taiyo Nippon Sanso Corp; Tokyo, Japan). SEs (1–2 mm length) were precultured for 1 day on MS solid medium with 0.3 M sucrose in the dark, followed by their transfer, one by one, to the wells of the cryoplates where they were encapsulated. First, embryos were covered with 2.5 μL of sodium alginate solution (2% w/v) containing 0.4 M sucrose, and then, with calcium chloride solution (0.1 M) gently added to allow polymerization of calcium alginate. After 15 min, calcium chloride solution was removed, and cryoplates with the encapsulated SEs were exposed to loading solution containing 1 M sucrose and 2 M glycerol for 15 min at room temperature, followed by the dehydration with a PVS solution pre-cooled in an ice bath. The effect of two vitrification solutions was evaluated: PVS2 (30% v/v glycerol, 15% v/v dimethyl sulfoxide, 15% v/v ethylene glycol, and 13.7% w/v sucrose) [24] and PVS3 (50% v/v glycerol and 50% w/v sucrose) for 0, 15, 30, 45, and 65 min prior (−LN) and after (+LN) direct immersion in LN [66]. Warming took place at room temperature using liquid MS as recovery medium with 1.2 M sucrose, where the cryoplates with samples were immersed for 15 min. Subsequently, the calcium alginate gel remaining in the SEs was carefully removed before the recovery culture. Post-cryopreservation recovery took place by transferring the embryos to semisolid MS medium supplemented with 0.3 M sucrose and culturing them for seven days in darkness at 25 °C, followed by the reculture onto semisolid MS medium and exposure to photoperiod. After cryopreservation, regrowth of SEs was evaluated after 45 days of culturing on MS medium, expressed by the elongation of coleoptile and the formation of radicles. The conversion rate of cryopreserved SEs to plants was determined after additional 60 days of culture in the MS medium (a total culture period of 105 d), and the morphological development of the obtained plants was compared with that of plantlets derived of non-cryopreserved SE and of ZE germinated in vitro. Non-cryopreserved controls were cultivated for 60 days in MS medium for the comparison.

4.4. Statistical Analysis

The experimental design was completely randomized to study both the embryogenesis and cryopreservation processes. Somatic embryogenesis experiments were replicated four times using four explants per Petri dish, and the results were expressed as the average ± standard deviations.

Cryopreservation experiments were replicated three times using ten SEs per replicate. Dependent variables were regrowth and the conversion rate of SEs to plant before ($-$LN) and after (+LN) the immersion in LN.

Results of somatic embryogenesis assays were analyzed by two-way analysis of variance (ANOVA) (levels of picloram and BAP). Results of cryopreservation assays were processed by one-way ANOVA (PVS). Means were compared by the least significant difference (LSD) range test with an error rate at $p \leq 0.05$. All statistical analyses were carried out using the Minitab® statistical software 17.2.1.

5. Conclusions

In this work, we reported the efficient application of in vitro techniques to induce indirect regeneration of SEs from *Agave tequilana* cultivar 'Chato' and their subsequent cryopreservation. The addition of growth regulators picloram and BAP to MS semisolid medium proved to be useful to induce somatic embryogenesis and obtain large amounts of actively growing embryos from callus derived of in vitro cultured leaf explants. The histological analysis illustrated this process and supported the single-cellular origin of SEs. The V-cryoplate method resulted in a practical and effective approach to cryopreserve SEs of cultivar 'Chato' using two PVS (PVS2 or PVS3). The experimental findings reported here represent viable alternatives to generate and safely store material for the long-term, which can be a new source of material for the commercial propagation of this plant species, the production of elite lines, and of usefulness for genetic transformation programs. This work provides new advances about somatic embryogenesis in *Agave* spp. and reports the first results on cryopreservation of SE of this species.

Author Contributions: Conceptualization, L.D.-A., M.T.G.-A. and L.P.; methodology, L.D.-A.; validation, L.D.-A., M.T.G.-A. and L.P.; investigation, L.D.-A. and L.P.; resources, L.D.-A., R.F. and L.P.; data curation, L.D.-A., M.T.G.-A.; writing—original draft preparation L.D.-A. and M.T.G.-A.; writing—review and editing, L.DA., M.T.G.-A., F.S.-R., R.F. and L.P.; visualization, M.T.G.-A.; supervision, F.S.-R. and L.P.; project administration, L.D.-A. and L.P.; funding acquisition, L.D.-A. All authors have read and agreed to the published version of the manuscript.

Funding: This work was funded by The Rufford Foundation (Grant 26560-1) and Consejo Nacional de Ciencia y Tecnología (Scholarship 291236).

Institutional Review Board Statement: Not applicable.

Informed Consent Statement: Not applicable.

Data Availability Statement: Not applicable.

Conflicts of Interest: The authors declare no conflict of interest. The funders had no role in the design of the study; in the collection, analyses, or interpretation of data; in the writing of the manuscript, or in the decision to publish the results.

References

1. Colunga-GarcíaMarín, P.; Zizumbo-Villarreal, D. Tequila and other agave spirits from west-central México: Current germplasm diversity, conservation and origin. *Biodivers. Conserv.* **2006**, *16*, 1653–1667. [CrossRef]
2. Palomino, G.; Dolezel, J.; Méndez, I.; Rubluo, A. Nuclear genome size analysis of *Agave tequilana* Weber. *Caryologia* **2003**, *56*, 37–46. [CrossRef]
3. Vargas-Ponce, O.; Zizumbo-Villarreal, D.; Martínez-Castillo, J.; Coello-Coello, J.; Colunga-Garciamarin, P. Diversity and structure of landraces of *Agave* grown for spirits under traditional agriculture: A comparison with wild populations of *A. angustifolia* (Agavaceae) and commercial plantations of *A. tequilana*. *Am. J. Bot.* **2009**, *96*, 448–457. [CrossRef] [PubMed]
4. Delgado-Lemus, A.; Torres-García, I.; Vázquez, J.; Alejandro, C. Vulnerability and risk management of *Agave* species in the Tehuacán Valley, México. *J. Ethnobiol. Ethnomed.* **2014**, *10*, 53. [CrossRef] [PubMed]
5. Zizumbo-Villarreal, D.; Vargas-Ponce, O.; Rosales-Adame, J.; Colunga-GarcíaMarín, P. Sustainability of the traditional management of *Agave* genetic resources in the elaboration of mescal and tequila spirits in western Mexico. *Genet. Resour. Crop Evol.* **2013**, *60*, 33–47. [CrossRef]
6. Valenzuela, A. A new agenda for blue agave landraces: Food, energy and tequila. *GCB Bioenergy* **2011**, *3*, 15–24. [CrossRef]
7. Diario Oficial de la Federación NOM-006-SCFI-2012 Proyecto de Norma Oficial Mexicana, Bebidas Alcohólicas-Tequila-Especificaciones. Available online: http://www.dof.gob.mx/nota_detalle.php?codigo=5282165&fecha=13/12/2012 (accessed on 8 June 2020).
8. Rodríguez, F.; Martínez, L.; Palomera, C. Contextualización socioambiental del agave en Tonaya, Jalisco, México. *Región Soc.* **2017**, *29*, 71–102. [CrossRef]
9. Powers, D.E.; Backhaus, R.A. *In vitro* propagation of *Agave arizonica* Gentry and Weber. *Plant Cell Tissue Organ Cult.* **1989**, *16*, 57–60. [CrossRef]
10. Santacruz-Ruvalcaba, F.; Gutiérrez-Pulido, H.; Rodríguez-Garay, B. Efficient in vitro propagation of *Agave parrasana* Berger. *Plant Cell Tissue Organ Cult.* **1999**, *56*, 163–167. [CrossRef]
11. Hazra, S.H.; Das, S.; Das, A.K. Sisal plant regeneration via organogenesis. *Plant Cell Tissue Organ Cult.* **2002**, *70*, 235–240. [CrossRef]
12. Martínez-Palacios, A.; Ortega-Larrocea, M.P.; Chavez, V.M. Somatic embryogenesis and organogenesis of *Agave victoria-reginae*: Considerations for its conservation. *Plant Cell Tissue Organ Cult.* **2003**, *74*, 135–142. [CrossRef]
13. Nikam, T.D.; Bansude, G.M.; Aneesh-Kumar, K.C. Somatic embryogenesis in sisal (*Agave sisalana* Perr. ex. Engelm). *Plant Cell Rep.* **2003**, *22*, 188–194. [CrossRef] [PubMed]
14. Portillo, L.; Santacruz-Ruvalcaba, F.; Gutiérrez-Mora, A.; Rodríguez-Garay, B. Somatic embryogenesis in *Agave tequilana* Weber cultivar azul. *Vitro Cell. Dev. Biol. Plant* **2007**, *43*, 569–575. [CrossRef]
15. Ramírez-Malagón, R.; Borodanenko, A.; Pérez-Moreno, L.; Salas-Araiza, M.D.; Nuñez-Palenius, H.G.; Ochoa-Alejo, N. *In vitro* propagation of three *Agave* species used for liquor distillation and three for landscape. *Plant Cell Tissue Organ Cult.* **2008**, *94*, 201–207. [CrossRef]
16. Torres, I.; Casas, A.; Vega, E.; Martínez-Ramos, M.; Delgado-Lemus, A. Population dynamics and sustainable management of mescal agaves in central Mexico: *Agave potatorum* in the Tehuacán Cuicatlán valley. *Econ. Bot.* **2015**, *20*, 1–16. [CrossRef]
17. Arzate-Fernández, A.; Piña-Escutia, J.L.; Norman-Mondragón, T.H.; Reyes-Díaz, J.I.; Guevara-Suárez, K.L.; Vázquez-García, L.M. Regeneration of agave (*Agave angustifolia* Haw.) from encapsulated somatic embryos. *Rev. Fitotec. Mex.* **2016**, *39*, 359–366.
18. Carneiro, F.; Queiroz, S.R.; Passos, A.; Nascimento, M.; Santos, K. Embriogênese somática em *Agave sisalana* Perrine: Indução, caracterização anatômica e regeneração. *Pesqui. Agropecuária Trop.* **2014**, *44*, 294–303. [CrossRef]
19. González-Arnao, M.T.; Martínez-Montero, M.; Cruz-Cruz, C.A.; Engelmann, F. Advances in cryogenic techniques for the long-term preservation of plant biodiversity. In *Biotechnology and Biodiversit*, 1st ed.; Ahuja, M.R., Ramawat, K.G., Eds.; Springer: Rajasthan, India, 2014; Volume 4, pp. 129–170.
20. Panis, B.; Withers, L.A.; De Langhe, E. Cryopreservation of *Musa* suspension cultures and subsequent regeneration of plants. *CryoLetters* **1990**, *11*, 337–350.
21. Lambardi, M.; De Carlo, A.; Capuana, M. Cryopreservation of embryogenic callus of *Aesculus hippocastanum* L. by vitrification/one-step freezing. *CryoLetters* **2005**, *26*, 185–192.
22. Yamamoto, S.; Rafique, T.; Priyantha, W.; Fukui, K.; Matsumoto, T.; Niino, T. Development of a cryopreservation procedure using aluminium cryo-plates. *CryoLetters* **2011**, *32*, 256–265.
23. Niino, T.; Yamamoto, S.; Fukui, K.; Castillo-Martínez, C.R.; Valle, M.; Matsumoto, T.; Engelmann, F. Dehydration improves cryopreservation of mat rush (*Juncus decipiens* Nakai) basal stem buds on cryo-plates. *CryoLetters* **2013**, *34*, 549–560. [PubMed]
24. Sakai, A.; Kobayashi, S.; Oiyama, I. Cryopreservation of nucellar cells of navel orange (*Citrus sinensis* Osb. var. brasiliensis Tnaka) by vitrification. *Plant Cell Rep.* **1990**, *9*, 30–33. [CrossRef] [PubMed]
25. Sakai, A.; Engelmann, F. Vitrification, encapsulation-vitrification and droplet-vitrification: A review. *CryoLetters* **2007**, *28*, 151–172. [PubMed]
26. Sánchez-Romero, C.; Sweenen, R.; Panis, B. Cryopreservation of olive embryogenic cultures. *CryoLetters* **2009**, *30*, 359–372. [PubMed]
27. Adu-Gyamfi, R.; Wetten, A. Cryopreservation of cocoa (*Theobroma cacao* L.) somatic embryos by vitrification. *CryoLetters* **2012**, *33*, 494–505.

28. Guzmán-García, E.; Bradaï, F.; Sánchez-Romero, C. Cryopreservation of avocado embryogenic cultures using droplet-vitrification method. *Acta Physiol. Plant* **2013**, *35*, 183–193. [CrossRef]
29. Jiménez, V.M. Regulation of in vitro somatic embryogenesis with emphasis on the role of endogenous hormones. *Rev. Bras. Fisiol. Veg.* **2001**, *13*, 196–223. [CrossRef]
30. Freire-Seijo, M. Aspectos básicos de la embriogénesis somática. *Biotecnol. Veg.* **2003**, *3*, 195–209.
31. Namasivayam, P. Acquisition of embryogenic competence during somatic embryogenesis. *Plant Cell Tissue Organ Cult.* **2007**, *90*, 1–8. [CrossRef]
32. Nhut, D.T.; Teixeira, J.A.; Aswath, C.R. The importance of the explant on regeneration in thin cell layer technology. *Vitro Cell. Dev. Biol. Plant* **2003**, *39*, 266–276. [CrossRef]
33. Portillo, L.; Santacruz-Ruvalcaba, F. Obtención de embrioides de *Agave tequilana* Weber a partir de explantes de raíz. *Zonas Áridas* **2006**, *10*, 11–19.
34. Monja-Mio, K.M.; Robert, M.L. Direct somatic embryogenesis of *Agave fourcroydes* Lem. through thin cell layer culture. *Vitro Cell. Dev. Biol. Plant* **2013**, *49*, 541–549. [CrossRef]
35. Naziri, M.; Sadat, S.; Soltani, M. The effect of different hormone combinations on direct and indirect somatic embryogenesis in *Agave americana. Plant Physiol.* **2018**, *9*, 2739–2747.
36. Rodríguez-Garay, B.; Gutiérrez-Mora, A.; Acosta-Dueñas, B. Somatic embryogenesis of *Agave victoria-reginae* Moore. *Plant Cell Tissue Organ Cult.* **1996**, *46*, 85–87. [CrossRef]
37. Tejavathi, D.H.; Rajanna, M.D.; Sowmya, R.; Gayathramma, K. Induction of somatic embryos from cultures of *Agave vera-cruz* Mill. *Vitro Cell. Dev. Biol. Plant* **2007**, *43*, 423–428. [CrossRef]
38. Flores-Benítez, S.; Jiménez-Bremont, J.F.; Rosales-Mendoza, S.; Argüello-Astorga, G.R.; Castillo-Collazo, R.A.; Alpuche-Solís, G. Genetic transformation of *Agave salmiana* by *Agrobacterium tumefaciens* and particle bombardment. *Plant Cell Tissue Organ Cult.* **2007**, *91*, 215–224. [CrossRef]
39. Alvarez-Aragón, C.; Arzate-Fernández, A.M.; Martínez-Martínez, S.Y.; Martínez-Velasco, I. Regeneration of *Agave marmorata* roezl plants, by somatic embriogénesis. *Trop. Subtrop. Agroecosyst.* **2020**, *23*, 1–13.
40. Santiz-Gómez, J.; Rincon-Rosales, R.; Gutiérrez-Miceli, F. In vitro propagation of *Agave grijalvensis* B.Ullrich, an endemic species from Chiapas under special protection. *Gayana Bot.* **2012**, *69*, 23–30.
41. Takamori, L.M.; Barbosa, M.N.; Esteves, V.L.; Ferreira, R.A. Optimization of somatic embryogenesis and *in vitro* plant regeneration of *Urochloa* species using picloram. *Vitro Cell. Dev. Biol. Plant* **2015**, *51*, 554–563. [CrossRef]
42. Beyl, C.A.; Sharma, G.C. Picloram induced somatic embryogenesis in *Gasteria* and *Haworthia. Plant Cell Tissue Organ Cult.* **1983**, *2*, 123–132. [CrossRef]
43. Williams, E.G.; Maheswaran, G. Somatic embryogenesis: Factors influencing coordinated behavior of cells as an embryogenic group. *Ann. Bot.* **1986**, *57*, 443–462. [CrossRef]
44. Ayala-González, C.; Gutiérrez-Mora, A.; Rodríguez-Garay, B. The occurrence of dicotyledonar embryos in *Agave tequilana. Biol. Plant* **2014**, *58*, 788–791. [CrossRef]
45. George, E.F. Plant Tissue Culture Procedure-Background. In *Plant Propagation by Tissue Culture*, 3rd ed.; George, E.F., Hall, M.A., De Klerk, G.J., Eds.; Springer: Dordrecht, The Netherlands, 2008; Volume 1, pp. 1–28.
46. Quiroz-Figueroa, F.R.; Fuentes-Cerda, C.F.J.; Rojas-Herrera, R.; Loyola-Vargas, V.M. Histological studies on the developmental stages and differentiation of two different somatic embryogenesis systems of *Coffea arabica. Plant Cell Rep.* **2002**, *20*, 1141–1149. [CrossRef]
47. Haccius, B. Question of unicellular origin of non-zygotic embryos in callus cultures. *Phytomorphology* **1977**, *28*, 74–81.
48. Halperin, W. In vitro embryogenesis: Some historical issues and unresolved problems. In *In Vitro Embryogenesis in Plants*, 1st ed.; Thorpe, T.A., Ed.; Kluwer Academic Publishers: Dordrecht, The Netherlands, 1995; Volume 20, pp. 1–16.
49. Dodeman, V.L.; Ducreux, G.; Kreis, M. Zygotic embryogenesis versus somatic embryogenesis. *J. Exp. Bot.* **1997**, *48*, 1493–1509.
50. Yamada, T.; Sakai, A.; Matsumura, T.; Higuchi, S. Cryopreservation of apical meristems of white clover (*Trifolium repens* L.) by vitrification. *Plant Sci.* **1991**, *78*, 81–87. [CrossRef]
51. González-Arnao, M.T.; Pants, A.; Roca, W.M.; Escobar, R.H.; Engelmann, F. Development and large-scale application of cryopreservation techniques for shoot and somatic embryo cultures of tropical crops. *Plant Cell Tissue Organ Cult.* **2008**, *92*, 1–13. [CrossRef]
52. González-Arnao, M.T.; Gámez, R.; Martínez, Y.; Valdés, S.; Mascorro, J.O.; Osorio, A.; Pastelín, M.; Guevara, M.; Cruz, C.A. Estado actual de la Crioconservación vegetal en México. In *Crioconservación de Plantas en América Latina y el Caribe*, 1st ed.; González-Arnao, M.T., Engelmann, F., Eds.; IICA: San José, Costa Rica, 2013; pp. 161–173.
53. San José, M.D.C.; Corredoira, E.; Oliveira, H.; Santos, C. Cryopreservation of somatic embryos of *Alnus glutinosa* (L.) Gaertn. and confirmation of ploidy stability by flow cytometry. *Plant Cell Tissue Organ Cult.* **2015**, *123*, 489–499. [CrossRef]
54. O'Brien, C.; Constantina, M.; Walia, A.; YuanYiing, J.; Mitter, N. Cryopreservation of somatic embryos for avocado germplasm conservation. *Sci. Hortic.* **2016**, *211*, 328–335. [CrossRef]
55. Bradaï, F.; Almagro-Bastante, J.; Sánchez-Romero, C. Cryopreservation of olive somatic embryos using the droplet-vitrification method: The importance of explant culture conditions. *Sci. Hortic.* **2017**, *218*, 14–22. [CrossRef]

56. Pettinelli, J.A.; Soares, B.O.; Collin, M.; Atalla, E.; Engelmann, F.; Gagliardi, R.F. Cryotolerance of somatic embryos of guinea (*Petiveria alliacea*) to V-cryoplate technique and histological analysis of their structural integrity. *Acta Physiol. Plant* **2020**, *42*, 1–10. [CrossRef]
57. Murashige, T.; Skoog, F. A revised medium for rapid growth and bioassays with tobacco tissue cultures. *Physiol. Plant* **1962**, *15*, 473–497. [CrossRef]
58. Phillips, G.C.; Collins, G.B. In vitro tissue culture of selected legumes and plant regeneration from callus cultures of red clover. *Crop Sci.* **1979**, *19*, 59–64. [CrossRef]
59. Castro-Concha, L.; Loyola-Vargas, V.M.; Chan, J.L.; Robert, M.L. Glutamate dehydrogenase activity in normal and vitrified plants of *Agave tequilana* Weber propagated in vitro. *Plant Cell Tissue Organ Cult.* **1990**, *22*, 147–151. [CrossRef]
60. Valenzuela-Sánchez, K.K.; Juárez, H.R.E.; Cruz, H.A.; Olalde, P.V.; Valverde, M.E.; Paredes, O. Plant regeneration of *Agave tequilana* by indirect organogenesis. *Vitro Cell. Dev. Biol. Plant* **2006**, *42*, 336–340. [CrossRef]
61. Von Arnold, S. Somatic Embryogenesis. In *Plant Propagation by Tissue Culture*, 3rd ed.; George, E.F., Hall, M.A., De Klerk, G.J., Eds.; Springer: Dordrecht, The Netherlands, 2008; Volume 1, pp. 335–354.
62. Quiroz-Figueroa, F.R.; Rojas-Herrera, R.; Galaz-Avalos, R.; Loyola-Vargas, V.M. Embryo production through somatic embryogenesis can be used to study cell differentiation in plants. *Plant Cell Tissue Organ Cult.* **2006**, *86*, 285–301. [CrossRef]
63. Burger, L.M.; Richter, H.G. *Anatomia da Madeira*, 1st ed.; Livraria Nobel: São Paulo, Brazil, 1991; pp. 97–105.
64. Gupta, P.; Durzan, D. Biotechnology of somatic polyembryogenesis and plantlet regeneration in loblolly pine. *Bio/Technology* **1987**, *5*, 147–151. [CrossRef]
65. Yamamoto, S.; Rafique, T.; Fukui, K.; Sekizawa, K.; Niino, T. V-Cryo-plate procedure as an effective protocol for cryobanks: Case study of mint cryopreservation. *CryoLetters* **2012**, *33*, 12–23.
66. Nishizawa, S.; Sakai, A.; Amano, Y.; Matsuzawa, T. Cryopreservation of asparagus (*Asparagus officinalis* L.) embryogenic suspension cells and subsequent plant regeneration by vitrification. *Plant Sci.* **1993**, *91*, 67–73. [CrossRef]

Article

Secondary Somatic Embryogenesis in *Centaurium erythraea* Rafn

Milica D. Bogdanović *, Katarina B. Ćuković, Angelina R. Subotić, Milan B. Dragićević , Ana D. Simonović, Biljana K. Filipović and Slađana I. Todorović

Department of Plant Physiology, Institute for Biological Research "Siniša Stanković"-National Institute of Republic of Serbia, University of Belgrade, Bulevar Despota Stefana 142, 11060 Belgrade, Serbia; katarina.cukovic@ibiss.bg.ac.rs (K.B.Ć.); heroina@ibiss.bg.ac.rs (A.R.S.); mdragicevic@ibiss.bg.ac.rs (M.B.D.); ana.simonovic@ibiss.bg.ac.rs (A.D.S.); biljana.nikolic@ibiss.bg.ac.rs (B.K.F.); slatod@ibiss.bg.ac.rs (S.I.T.)
* Correspondence: milica.bogdanovic@ibiss.bg.ac.rs

Abstract: Somatic embryogenesis (SE) is a developmental process during which plant somatic cells, under suitable conditions, produce embryogenic cells that develop into somatic embryos (**se**). SE is the most important method for plant propagation in vitro, having both fundamental and applicative significance. SE can be induced from different tissues and organs, but when **se** are used as explants, the process is recognized as secondary or cyclic SE. We induced secondary SE in *Centaurium erythraea* by application of 2,4-dichlorophenoxyacetic acid (2,4-D) and *N*-(2-chloro-4-pyridyl)-*N'*-phenylurea (CPPU). A medium containing 0.1 mgL^{-1} 2,4-D and 0.25 mgL^{-1} CPPU was optimal in terms of the number of primary SE explants forming **se**, the number of well-developed **se** per explant, and morphological appearance of the obtained **se**. These concentrations allowed SE to progress through three cycles, whereas at higher concentrations of 0.2 mgL^{-1} 2,4-D and 0.5 mgL^{-1} CPPU, only two cycles were achieved. Histological analysis revealed that secondary **se** are formed both directly and indirectly. Secondary SE readily germinated and converted into plantlets. Induction of cyclic SE contributes to the conservation efforts of this endangered medicinal plant and expands the spectrum of in vitro developmental pathways described in centaury—an emerging model in developmental biology.

Keywords: cyclic somatic embryogenesis; direct somatic embryogenesis; indirect somatic embryogenesis; leaf explant; histology; 2,4-D; CPPU; auxins; cytokinins

Citation: Bogdanović, M.D.; Ćuković, K.B.; Subotić, A.R.; Dragićević, M.B.; Simonović, A.D.; Filipović, B.K.; Todorović, S.I. Secondary Somatic Embryogenesis in *Centaurium erythraea* Rafn. *Plants* **2021**, *10*, 199. https://doi.org/10.3390/plants10020199

Received: 29 December 2020
Accepted: 14 January 2021
Published: 21 January 2021

Publisher's Note: MDPI stays neutral with regard to jurisdictional claims in published maps and institutional affiliations.

1. Introduction

Centaurium erythraea Rafn (common, European, or small centaury), belonging to the Gentianaceae family, is a medicinal plant with a broad environmental tolerance. Centaury is widespread over most of Europe, where it grows in different habitats, such as on river banks and wood margins, as well as on calcareous dry and sandy land [1,2]. The aerial part of the plant, *Centaurii herba*, is traditionally used as bitter tinctures, tonics, lotions, or teas to treat a diversity of ailments. The bitter taste is due to the secoiridoids. Secoiridoid glucosides are reported in various applications for the treatment of different digestive problems, as well as gastroprotective [3] and hepatoprotective agents [4]. Other important secondary metabolites include xanthones [5], as well as alkaloids, terpenoids, phenolic acids, flavonoids, fatty acids, alkanes, and waxes [6,7], some of which are constituents of centaury essential oils [8]. Centaury also exhibits considerable antioxidant [7,9,10], antidiabetic [11,12], and antimicrobial [7] pharmacological properties, which are correlated with its phytochemical composition. Because of the extensive and uncontrolled exploitation, coupled with its limited cultivation restricted by unpredictable seed germination and the inability of *C. erythraea* to grow in dense stands [6], as well as insufficient attempts for the replenishment, the wild populations of centaury have been markedly depleted. Sustainable utilization of this valuable medicinal plant and the efforts for its conservation, as well

73

as biotechnological alternatives for the production of its secondary metabolites, rely on the development of efficient in vitro techniques for the mass propagation of centaury [13]. As reviewed in the accompanying article in this issue [14], the most extensively studied pathway of centaury in vitro propagation is somatic embryogenesis (SE).

SE is a developmental process by which plant somatic cells, under suitable inductive conditions, produce embryogenic cells that, through a series of morphological and biochemical changes, form a somatic embryo [15]. Somatic embryo (**se**) is a structure that resembles the zygotic embryo, but formed without fertilization, which passes through similar stages (globular, heart-shaped, torpedo-shaped, and cotyledonary embryo). As such, **se** is not enclosed by maternal tissues, so that the process of SE can be not only controlled by the in vitro culture conditions, but the obtained **se** can be observed and collected at different developmental stages for the molecular and biochemical analyses [15]. Thus, SE is used as a model system for studying morphological, physiological, and molecular aspects of embryogenesis in higher plants [15,16], as well as for investigating cellular differentiation and mechanisms leading to acquisition of totipotency in plant cells [17]. Equally important are various biotechnological applications of SE, such as the propagation of elite or transgenic lines, while single-cell origin of some **se** may offer many advantages for breeding programs [18]. Actually, SE is considered as the most appropriate in vitro method for the clonal propagation of different plant species due to its high multiplication potential [19]. In addition, plant propagation trough SE represents an important source of material for plant transformation, offering genetically identical starting material, with less somaclonal variation as compared to propagation through organogenesis [20]. SE has also been revealed as the best regeneration pathway in cryopreservation [21], as well as a method of choice for the haploid production, somatic hybridization, and the production of artificial seeds [16].

Somatic embryos can either differentiate directly, from a small group of cells of the explanted tissue (direct SE or DSE), or indirectly, from embryogenic callus cells that further produce embryos (indirect SE or ISE) [22]. SE is influenced by internal and external factors, such as the type and the physiological state of the explant, the composition of the culture medium, the type and concentration of plant growth regulators (PGRs) in the medium, temperature, and light regime [23]. It has been suggested that in the case of DSE, proembryogenic competent cells are already present in the explant, hence they require minimal reprogramming, while in ISE, major cell reprogramming is needed to acquire embryogenic potential [24].

In most plant species, particularly in the Gentianaceae family, PGRs in the auxin and cytokinin groups are among the main factors affecting the induction of SE [25]. They determine the acquisition of totipotency by the explant cells [26], and consequently induce the development of **se**. Auxins and cytokinins are involved in the regulation of cell divisions and differentiation processes in the plant tissues [27], leading to the formation of SE. In the protocols for the induction of SE in many plant species, cell reprogramming is induced by a treatment with exogenous auxin, usually 2,4-dichlorophenoxyacetic acid (2,4-D). The evidence supports the notion that auxins play a critical role in cell reprogramming, while the induction of SE development requires subsequent elimination of the auxin from the culture media [28]. The concentrations of auxin required for SE induction may vary with different protocols. Cytokinins are thought to be more involved in the differentiation and further development of SE. In most plant species, the SE induction requires the presence of both PGRs [29], but SE can also be achieved by using only certain auxins [30] or cytokinins [31].

Secondary SE is a developmental process by which new (secondary) somatic embryos are developed from the primary somatic embryos used as explants. Other common names for secondary SE are repetitive, cyclic, recurrent, accessory, or adventitious SE. Secondary **se** are formed directly or/and indirectly on the cotyledons, hypocotyls, epicotyls, or root tips of the primary **se**. Efficient plant regeneration systems through secondary SE have been reported in several plant species, for example in *Cyclamen persicum* [32], *Hepatica nobilis* [33], *Pseudotsuga menziesii* [34], *Akebia trifoliata* [35], and *Olea europea* [36]. Secondary SE may enhance and prolong embryogenic competence of certain lines [34], multiply the number of embryos that can be obtained (compared to primary SE) [34], and recycle **se** of abnormal morphology that otherwise cannot regenerate normal plants [37]. This developmental pathway characterizes high multiplication index, repeatability, independence from explants source effects, and high level of uniformity [38]. Low production rate of important clone lines can be enhanced by obtaining secondary SE [34]. Since in many species, embryogenic competence in the in vitro culture declines over time due to aging and subculturing for several months [39,40], secondary SE provides a way to restore the embryogenic potential of important productive lines and is routinely used with broadleaved tree species as a method of long-term management [34].

As discussed in detail in the accompanying review article of this issue [14], several successful protocols for the induction of SE in *C. erythraea* from different explants have been published. Briefly, SE has been induced in cell suspension culture [41], root explants [6,42–45], and leaf explants [13,46]. While SE from roots was spontaneous and direct, the SE from the leaf explants was indirect and induced by the addition of 2,4-D and a urea-type synthetic cytokinin *N*-(2-chloro-4-pyridyl)-*N'*-phenylurea (CPPU). There are no reports on the induction of secondary SE in centaury. Hereby, we report a successful establishment of secondary SE in centaury, as a valuable addition to the spectrum of protocols for the in vitro propagation of this species. Namely, we have recently proposed *C. erythraea* as a model plant organism in developmental biology due to its great regeneration potential and developmental plasticity [46,47]. So, the present work aims not only to provide a more efficient way for the in vitro propagation of centaury as a support for the conservation efforts, but also to establish an additional system for fundamental research of centaury development. Namely, having several systems for the induction of SE from different explants in the same species would allow for a comparison of biochemical and molecular events in these systems within the same genetic background. The effects of different combinations of 2,4-D and CPPU on the induction of SE are described, along with the effects of multiple cycles of SE on the efficiency of this process. The results are supported by histological analyzes of the embryogenic tissues and developing **se**, as well as the germination tests of the obtained secondary **se**.

2. Results

2.1. Induction of Primary SE from C. erythraea Leaf Explants and the Experimental Setup

We have previously published successful induction of SE from centaury leaf explants cultivated on a combination of 2,4-D and CPPU, where ISE proceeds as a sole developmental pathway, providing that the leaf explants are kept in darkness [13]. Thus, primary SE was induced as described by Filipović et al. [13], with slight modifications. The leaf segments of mature plants were cultivated for three weeks in the darkness on MS medium supplemented with 0.1 mgL^{-1} 2,4-D and 0.25 mgL^{-1} CPPU (Figure 1). Well-developed primary cotyledonary somatic embryos (**cse**) (Figures 1 and 2a) formed on this media were used as primary explants for the induction of secondary SE on media with varying 2,4-D and CPPU concentrations (Figure 1). Generally, in all experiments described below, only **cse** were used as explants, even though embryos of all developmental stages were observable. For a comparison, the leaf explants were also cultivated on a media supplemented with 0.2 mgL^{-1} 2,4-D and 0.5 mgL^{-1} CPPU, and the primary **cse** obtained in this setup were also used for the induction of secondary and cyclic SE, as discussed later.

2.2. Induction of Secondary SE

Well-developed primary **cse** formed on the leaf explants on 0.1 mgL^{-1} 2,4-D and 0.25 mgL^{-1} CPPU (Figure 2a) were excised (Figures 2b and 3a) and used as explants for the induction of secondary SE on the same medium. After four weeks, primary **cse** explants enlarged several times and developed both embryogenic calli (**ec**) and nonembryogenic calli (**nec**), as well as somatic embryos at different developmental stages (Figure 2c). The **nec** that developed on primary **cse** was watery, friable, and translucent (Figure 3b). In contrast, the **ec** exhibited more organized structure and morphology: the embryogenic tissue was semi-compact, nodular, with a smooth surface and whitish to yellowish color (Figure 3c–e). Secondary **se** developed on the primary **cse** both indirectly, from the **ec** (Figure 3e,f,h), as well as directly on the explants, without callusing (Figure 3g,h). Somatic embryos at the cotyledonary stage formed directly or indirectly are referred to in the following text as **dcse** and **icse**, respectively. Although most of the explants swelled and significantly changed their morphology during the cultivation, in some of the explants, **dcse** could be observed developing on the primary cotyledons (Figure 3g).

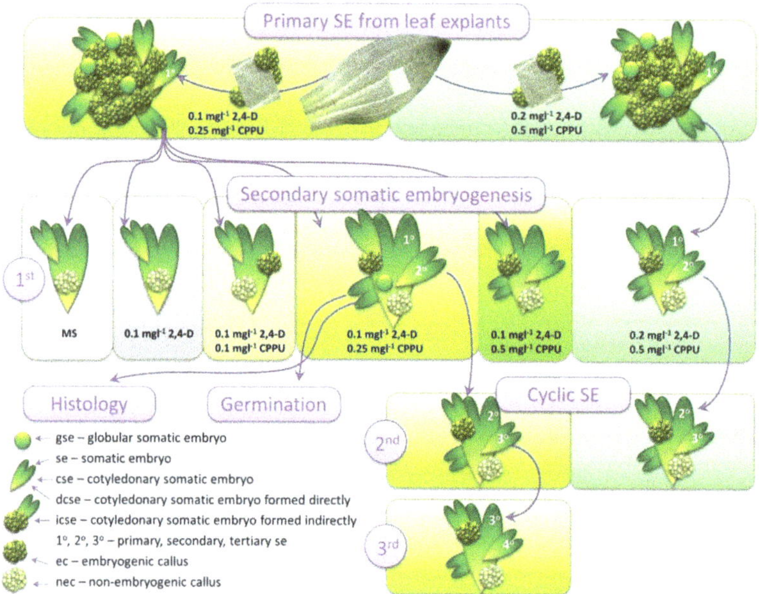

Figure 1. Schematic presentation of the experimental setup. Primary cotyledonary somatic embryos (**cse**) developed on the centaury leaf explants provided the initial explants for the induction of secondary somatic embryos (SE). The induction of cyclic SE was conducted on two different media. Secondary embryos developed on 0.1 mgL^{-1} 2,4-dichlorophenoxyacetic acid (2,4-D) and 0.25 mgL^{-1} *N*-(2-chloro-4-pyridyl)-*N'*-phenylurea (CPPU) were examined histologically and their germination was tested. Rectangles of identical colors represent the same composition of the culture medium. The listed abbreviations are used throughout the text.

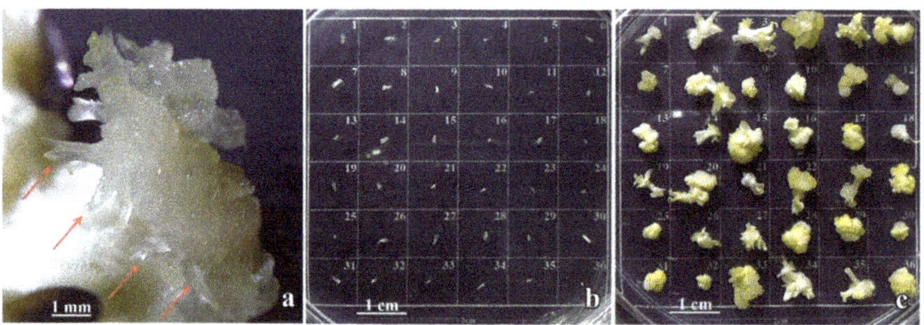

Figure 2. The induction of secondary SE in *Centaurium erythraea*. (**a**) Primary **cse** developed on leaf explants (red arrows); (**b**) primary **cse** were excised and set as explants arranged as 6 × 6 array on MS medium containing 0.1 mgL^{-1} 2,4-D and 0.25 mgL^{-1} CPPU; (**c**) primary **cse** with induced secondary embryos.

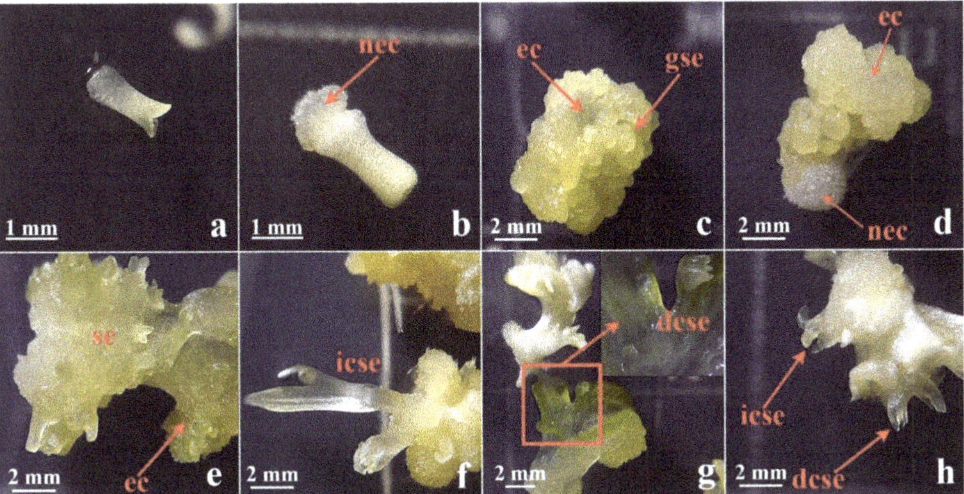

Figure 3. Secondary SE after four weeks in culture on medium with 0.1 mgL^{-1} 2,4-D and 0.25 mgL^{-1} CPPU. (**a**) Primary explant—**cse** at the beginning of the experiment; (**b**) **cse** with only nonembryogenic calli (**nec**) developed; (**c**) **cse** with developed embryogenic calli (**ec**) and globular somatic embryo (**gse**); (**d**) **cse** that developed both **ec** and **nec**; (**e**) **ec** with secondary **se** of various developmental stages; (**f**) cotyledonary somatic embryo formed indirectly (**icse**) developed from callus; (**g**) cotyledonary somatic embryo formed directly (**dcse**) developed on cotyledons of primary **cse** explants; (**h**) both **dcse** and **icse** types of embryos on the same explant. **cse**—cotyledonary somatic embryo, **nec**—nonembryonic callus, **ec**—embryogenic callus, **gse**—globular somatic embryo, **dsce**—cotyledonary somatic embryo formed directly, **icse**—cotyledonary somatic embryo formed indirectly.

2.3. Histology of the Secondary SE

The process of the development of secondary somatic embryos in centaury was histologically analyzed. As explained in the previous section, secondary SE was induced on media containing 0.1 mgL^{-1} 2,4-D and 0.25 mgL^{-1} CPPU. Morphological features of the explants with directly and/or indirectly developed secondary **se** at different stages, that were used for the histological analysis, are shown in Figure 4a–c. The presence of the **se** at different developmental stages that could be observed on the same explant suggests that the process of SE is asynchronous. The major events during the SE are comprised of early, intermediate, and the maturation stages. Early stages of SE are described as the process of the induction of **ec**, as well as induction of direct **se** from the subepidermal cells of the explant. This is followed by the intermediate stage of SE, during which **se** are initiated either indirectly, from the proembryogenic masses (PEM), or directly, following the activation of repeated cell divisions of the proembryogenic cells. Finally, the maturation stage of SE is the final stage of vascular patterning for the induction of shoot apical meristem (SEM), leaf primordia, and provascular bands. Histological features observed during DSE from centaury roots [42,43] and ISE from the leaf explants [13] aid in distinguishing these two types of SE, which occur simultaneously during secondary SE.

Direct induction of secondary **se** was seen from the entire surface of the primary **se**. The direct initiation of **se** was observed from the subepidermal layer of the **cse** explant without an intervening callus phase. Differentiated multicellular PEMs, seen at the periphery of the explants, further developed into embryos (Figure 4d). During further growth, the PEM produced secondary somatic embryos at the globular stages of development (**gse**). The **gse** at the onset of polarization, showing a protoderm-like layer, were the first clearly distinguishable stage of the somatic embryo differentiation (Figure 4e). These **gse** had no apparent vascular connection with the primary **cse** tissue to which they are connected by a suspensor-like structure (Figure 4e). Subsequent somatic embryo development included their elongation, development of procambium, and shoot meristem differentiation, reaching the early cotyledonary-staged somatic embryo (Figure 4f). The secondary **cse** were poorly attached to the surface of the primary explants and could easily separate. No vascular connections were observed between the developing somatic embryos and the primary explant (Figure 4f).

As observed using light microscopy, the **ec** was formed by small and isodiametric clumps of cells, containing prominent nuclei and dense cytoplasm. These clusters were round structures bounded by a layer of organized cells. Histological analyses showed that some cells in the superficial portions of this callus exhibited meristematic characteristics (Figure 4g). These clusters of proembryogenic cells progressed through a series of organized division to give rise to **gse** (Figure 4h). Finally, an increased embryo differentiation and elongation were observed, which became sharper, reaching the late **cse** with well-developed shoot apical meristem, leaf primordia and provascular bands (Figure 4i).

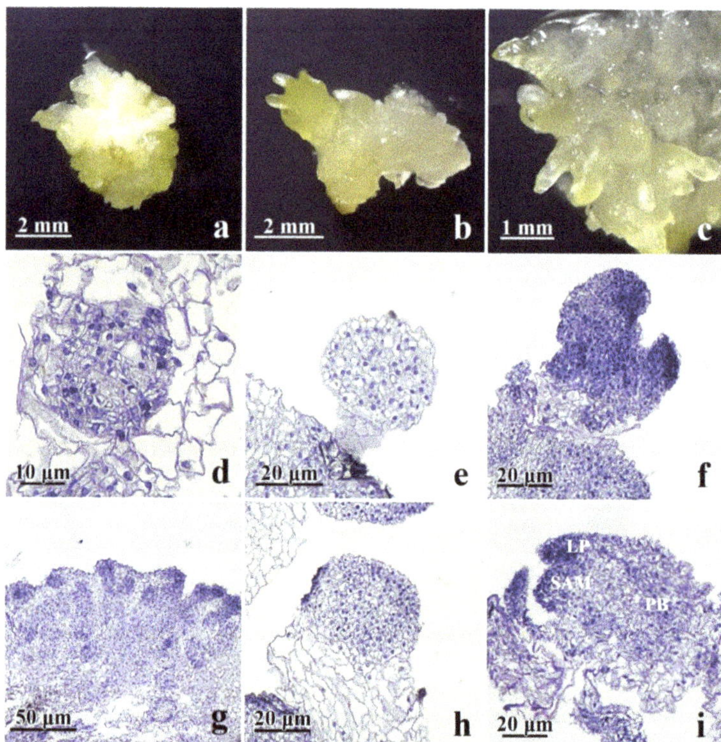

Figure 4. Morphological and histological characterization of secondary SE from primary **cse** explants cultured for 3 weeks in the darkness on medium supplemented with 0.1 mgL^{-1} 2,4-D and 0.25 mgL^{-1} CPPU. (**a–c**) Numerous secondary **cse** were visible to arise indirectly from **ec** and/or directly from the primary explant; (**d–f**) the process of secondary DSE; (**g–i**) the process of secondary ISE; (**d**) developed proembryogenic cell masses (PEMs); (**e**) formation of secondary **gse** with a suspensor-like structure at the surface of primary **cse**; (**f**) secondary **cse** at early stage of development; (**g**) embryogenic callus formed of small and isodiametric clumps of cells; (**h**) globular somatic embryo formed from **cse**; (**i**) secondary **cse** at a late developmental stage with visible shoot apical meristem (SAM), provascular bands (PB), and leaf primordia (LP).

2.4. The Effect of 2,4-D and CPPU on the Induction of Secondary SE

To explore the effect of different concentrations of 2,4-D and CPPU on the process of secondary SE, well-developed primary **cse** were excised from the leaf explants cultured on 0.1 mgL^{-1} 2,4-D and 0.25 mgL^{-1} CPPU and transferred to fresh media with different content of PGRs. The media were supplemented with 2,4-D (0.1 mgL^{-1}) and increasing CPPU concentrations (0–0.5 mgL^{-1}), whereas a hormone-free medium was used as control (Figures 1 and 5). After 4 weeks in culture, each primary **cse** was observed, and features such as development of calli (**ec** or **nec**) and **se** at different developmental stages (cotyledonary or other) and of different origin (directly or indirectly formed) were recorded as a number of explants with a particular feature.

General regenerative potential of the **cse** explants on different media was initially evaluated as the number of explants forming calli. On a hormone-free medium, on average, only 3 out of 36 explants formed calli, exclusively nonembryogenic ones (**nec**), while no explants formed embryogenic (**ec**) calli (Figure 5a). In the **cse** explants cultivated on 0.1 mgL^{-1} 2,4-D, on average, about 60% of the explants produced calli, but again, all of them were **nec**. However, if both PGRs were added to the medium, almost all of the **cse** (on average 34.14–35.57 explants per treatment) generated calli, even at the lowest CPPU concentration of 0.1 mgL^{-1} (Figure 5a). Therefore, CPPU was necessary for the generation of **ec**. The concentration of 0.5 mgL^{-1} CPPU gave the highest number of explants with **ec** (21.71 on average), with concomitant reduction of the number of **cse** where only **nec** formed (Figure 5a).

Even though the primary explants cultivated on media without CPPU did not produce any embryogenic calli, secondary **se** did appear, on average, on 10 (out of 36) explants on the MS medium, and on 6.57 explants grown on 0.1 mgL^{-1} 2,4-D (Figure 5b). Obviously, in the absence of **ec**, all of the secondary **se** formed on these media were developed directly. Spontaneous secondary DSE is depicted in Figure 6a. With the addition of CPPU, the number of explants developing **se** drastically increased in a dose–response manner, up to, on average, 35.14 explants on the media supplemented with 0.5 mgL^{-1} CPPU (Figure 5b). Only **cse** were further classified into embryos formed directly and indirectly (**dcse** and **icse**), and it turned out that DSE is not only the exclusive path on CPPU-free media, but also the predominant path on media containing CPPU (Figure 5b). However, the number of explants with embryos developed by ISE slightly increased with rising the CPPU concentration, up to an average of 5.43 explants on medium with 0.1 mgL^{-1} 2,4-D and 0.5 mgL^{-1} CPPU (Figure 5b, "**icse** only"), in concordance with higher induction of **ec** on this medium (Figure 5a).

In addition, the number of **cse** per explant was also scored after 4 weeks of cultivation on different media (Table 1). The average number of the **cse** per explant was significantly higher on media containing 0.25 and 0.5 mgL^{-1} CPPU, as compared to other media, reaching up to 3.87 ± 0.50 new secondary **cse** per primary explant (Table 1).

The induction of secondary SE on different media produced **cse** of different morphology. On the hormone-free media, secondary **se** formed on the primary explants spontaneously and directly, mostly in the hypocotyl zone of the explant, and had normal morphology (Figure 6a). Concentration of 0.1 mgL^{-1} 2,4-D with 0.25 mgL^{-1} CPPU resulted in a production of well-developed bi-cotyledonary **se** (Figure 6b), which were easily isolated from the primary explant (and thus preferable as explants for new cycles of SE, as discussed later). Secondary **cse** forming on media formulations with higher CPPU concentration (0.5 mgL^{-1} CPPU) often had an abnormal morphology, with **cse** having more than two cotyledons, trumpet or fascicular shape, and fused cotyledons (Figure 6c,d).

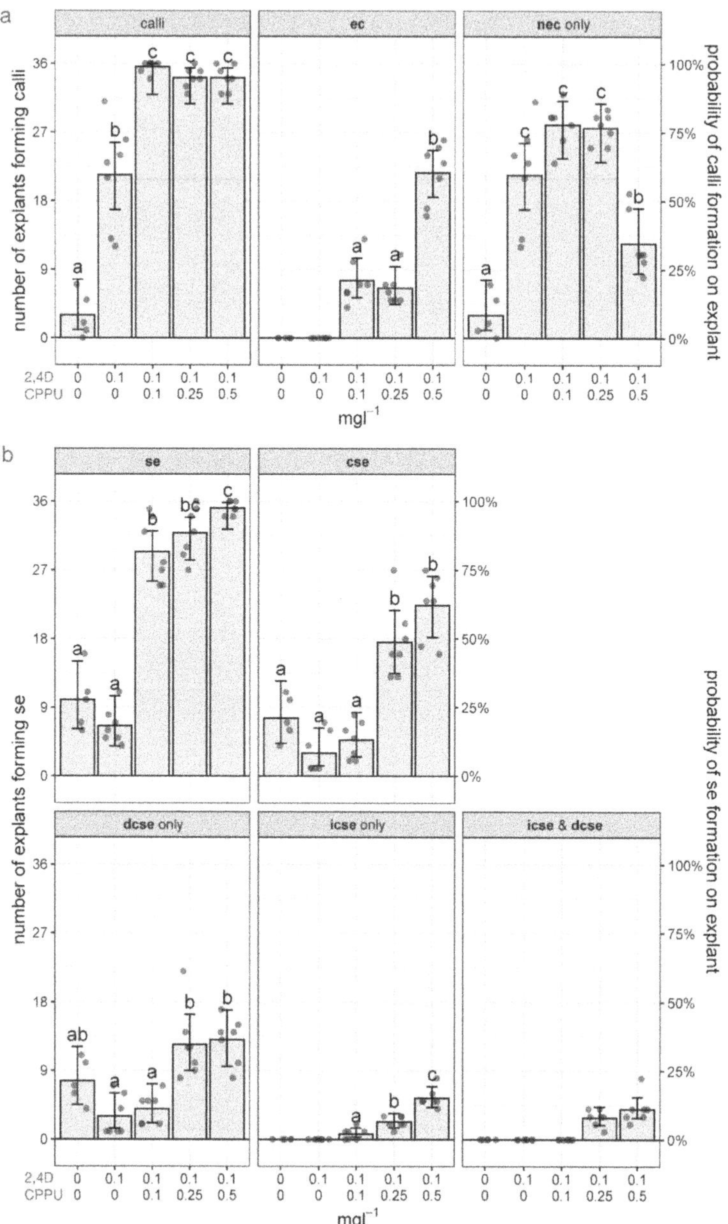

Figure 5. The effect of different concentrations of 2,4-D and CPPU on the induction of calli and secondary embryos on the **cse** explants. (**a**) The number of **cse** explants forming any type of calli and specifically embryogenic (**ec**) or nonembryogenic (**nec**) calli; (**b**) the number of **cse** explants forming secondary embryos of any stage or origin (**se**) and specifically cotyledonary somatic embryos (**cse**), formed directly (**dcse**) or indirectly (**icse**). The mean values for seven replicates (with 36 **cse** explants each) equivalent to fitted values of the corresponding logistic regression models, with error bars (95% confidence intervals) are presented as the number of explants forming specific types of calli or SE (left ordinate), or the probability of calli or SE formation on the explants (right ordinate). Different letters denote statistically significant differences at $p < 0.05$. Gray dots represent individual replicates (Petri dishes).

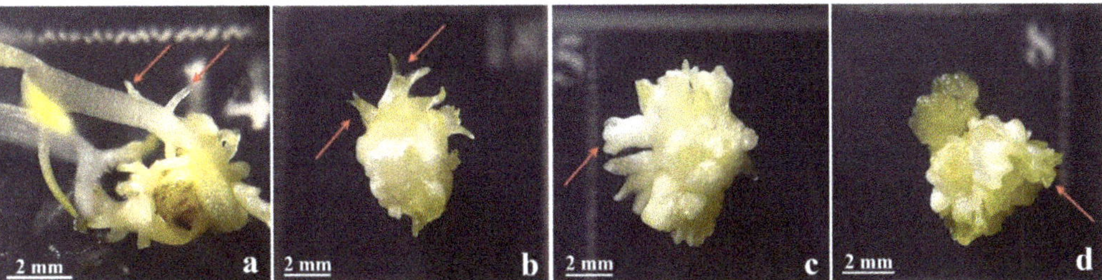

Figure 6. Morphology of secondary **cse** (red arrows) developing on different hormone combinations: (**a**) hormone-free medium; (**b**) 0.1 mgL^{-1} 2,4-D with 0.25 mgL^{-1} CPPU; (**c**) 0.1 mgL^{-1} 2,4-D with 0.5 mgL^{-1} CPPU; and (**d**) 0.2 mgL^{-1} 2,4-D with 0.5 mgL^{-1} CPPU.

Table 1. The effect of media formulation on the average number of secondary **cse** per primary explant. Only explants that formed **cse** were considered. Mean values ± SD per treatment are presented. Different letters denote significant differences ($p < 0.05$) according to Tukey post-hoc pairwise comparisons.

2,4-D [mg l^{-1}]	CPPU [mg l^{-1}]	cse Per Explant
0	0	1.43 ± 0.44 [a]
0.1	0	1.18 ± 0.37 [a]
0.1	0.1	1.61 ± 0.54 [a]
0.1	0.25	3.28 ± 0.71 [b]
0.1	0.5	3.87 ± 0.50 [b]

2.5. Induction of Cyclic SE

To see whether SE can continue through several cycles, two combinations of PGRs were tested: 0.1 mgL^{-1} 2,4-D and 0.25 mgL^{-1} CPPU or 0.2 mgL^{-1} 2,4-D and 0.5 mgL^{-1} CPPU, referred in the following text as lower and higher concentrations, respectively. These two combinations of PGRs were maintained throughout this experiment, starting with the induction of SE on the leaf explants (see Figure 1). The four replicates (Petri dishes) that were used in this experiment were linked through the cycles in terms that the embryos developed in one cycle were transferred to a Petri dish with the same label in the next cycle. Since the highest number of the explants forming **se** or **cse** (Figure 5b), as well as the highest number of the **cse** per explant (Table 1), were obtained on 0.1 mgL^{-1} 2,4-D and 0.25 or 0.5 mgL^{-1} CPPU, the lower concentration of CPPU was used in combination with 2,4-D (0.1 mgL^{-1} 2,4-D and 0.25 mgL^{-1} CPPU) because the embryos developed on this media exhibited normal morphology (Figure 6b). On the other hand, the higher concentrations of both PGR but at the same ratio (0.2 mgL^{-1} 2,4-D with 0.5 mgL^{-1} CPPU) proved to be optimal for the induction of SE in centaury leaf culture [13], and for this reason were also used in this experiment. Only somatic embryos at the cotyledonary stage (**cse**) were harvested and set to initiate the next cycle.

In the 1st cycle of the cyclic SE, secondary **cse** that developed on primary **cse** explants on media with lower or higher PGRs concentration were excised, transferred to a fresh media of the same composition, and their morphological features (developed calli and embryos) were scored after 4 weeks in culture. In the 2nd cycle, tertiary **cse** that formed after four weeks were subcultured on the same fresh media for the 3rd and last cycle (see Figure 1). As can be seen in Figure 7, both hormone combinations were able to initiate cyclic SE, but as discussed later, only SE on the lower concentrations could progress through three cycles, whereas only two cycles were possible on higher concentrations.

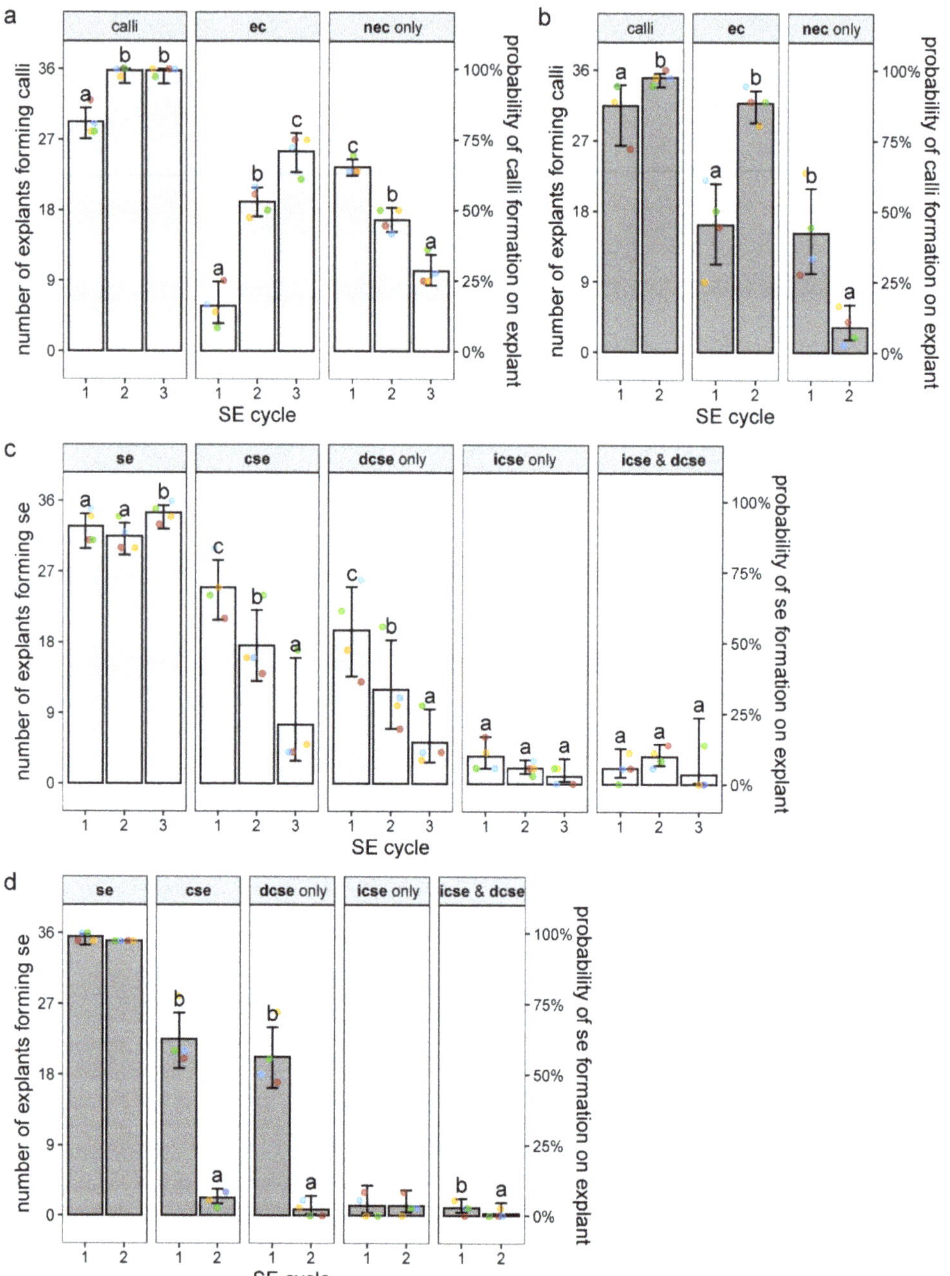

Figure 7. The effect of secondary SE cycle and media composition on the induction of calli and secondary embryos on explants. (**a**,**c**) The explants cultivated on lower concentrations of plant growth regulators (PGRs) (0.1 mgL^{-1} 2,4-D and 0.25 mgL^{-1} CPPU); (**b**,**d**) the explants are cultivated on higher concentrations of PGRs (0.2 mgL^{-1} 2,4-D and 0.5 mgL^{-1} CPPU); (**a**,**b**) the number of **cse** explants forming any type of calli and specifically embryogenic (**ec**) or nonembryogenic (**nec**) calli; (**c**,**d**) the number of **cse** explants forming secondary embryos of any stage or origin (**se**) and specifically cotyledonary

somatic embryos (**cse**), formed directly (**dcse**) or indirectly (**icse**). The mean values for four replicates (with 36 **cse** explants each), equivalent to fitted values of the corresponding general estimating equations models, with error bars (95% confidence intervals), are presented as the number of explants forming specific types of calli or **se** (left ordinate), or the probability of calli or **se** formation on the explants (right ordinate). Dots of the same color represent linked replicates (Petri dishes) through the cycles. Different letters denote statistically significant differences at $p < 0.05$.

On the medium with lower PGRs, on average, 29.25 of the **cse** explants produced some type of calli during the 1st cycle, whereas nearly all explants (35.75 on average) generated calli during the 2nd and the 3rd cycle (Figure 7a). Likewise, the number of explants forming calli increased in the 2nd cycle, as compared to the 1st cycle, on higher PGRs concentration (Figure 7b). The embryogenic capacity of the explants measured as the number of **cse** explants producing embryogenic calli (**ec**) also significantly increased with the cycles' progression on both types of media. Specifically, the number of explants forming **ec** on 0.1 mgL^{-1} 2,4-D and 0.25 mgL^{-1} CPPU increased through the cycles from 5.75 explants (per replicate) in the 1st cycle to 25.5 explants in the 3rd cycle (Figure 7a), while on 0.2 mgL^{-1} 2,4-D and 0.5 mgL^{-1} CPPU, 16.25 and 31.75 explants produced **ec** in the 1st and 2nd cycle, respectively (Figure 7b). In both cases, this increase was concomitant with a decrease in the number of explants forming only **nec** (Figure 7a,b).

The formation of **se**, as another parameter of the embryogenic capacity of the explants, also did not change much over the cycles, since nearly all of the explants formed **se** on both types of media (Figure 7c,d). On the lower PGRs concentration, there was a slight (albeit statistically significant) increase in the number of explants forming **se** in the 3rd cycle, as compared to the first two cycles (Figure 7c), whereas on higher concentration, nearly all of the explants developed **se** in both cycles (Figure 7d). However, the number of fully-developed embryos at the cotyledonary stage drastically decreased with the cycles' progression at both lower and higher concentrations of PGRs. Thus, the number of explants with developed **cse** decreased from 25 in the 1st to 7.5 in the 3rd cycle, and from 22.5 in the 1st to only 2.25 in the 2nd cycle on lower and higher concentrations, respectively (Figure 7c,d). Considering only explants that did form **cse**, the average number of newly-formed **cse** per explant was not significantly different among different SE cycles and different concentrations, being ≈3 in all cases (data not shown). This means that in the 2nd cycle at the higher concentration, less than 10 **cse** were formed, which was insufficient to initiate the 3rd cycle. Overall, 432 **cse** that were used as explants on the lower concentrations (4 Petri dishes x 36 explants/replicate x 3 cycles) produced a total of 634 newly-formed, well-developed **cse** in all cycles, whereas 288 **cse** cultivated on higher PGRs concentrations (4 replicates x 36 explants x 2 cycles) formed a total of 280 **cse**. Most of the **cse** were formed by direct pathway on both media and in all cycles (Figure 7c,d).

2.6. Germination of Secondary Somatic Embryos

The ability of secondary **cse**, obtained on media with 0.1 mgL^{-1} 2,4-D and 0.25 mgL^{-1} CPPU, to germinate and develop into plantlets was tested in the light and in darkness on media without growth regulators. In the light, well-developed shoots with multiple leaves were formed on 80% of the explants (Figure 8a–c), while 32.4% of these plantlets developed roots after 25 days of cultivation (Figure 8b,c). Well-developed and rooted plantlets (Figure 8e) further developed into plants upon transfer to fresh media (Figure 8f). Most plants appeared to be healthy and to grow vigorously. In darkness, 64.5% of the **cse** explants developed etiolated shoots (Figure 8d), and most of those shoots also formed roots.

Figure 8. Germination of secondary **cse** on media without PGRs. (**a–c**) Most of secondary **cse** germinated in the light had well-developed shoots and some of them formed roots; (**d**) when the secondary **cse** are set to germinate in the darkness, roots and etiolated shoots are formed; (**e**) 40 days old plantlet with well-developed roots; (**f**) plantlets ready for transfer to MS for further development.

3. Discussion

One of the most important factors in SE induction in most plant species is the concentration of auxins and cytokinins present in the medium. Regarding the requirements for PGRs for the induction of secondary SE, Raemakers et al. [18] concluded that the kinds of PGRs suitable for primary SE were generally suitable for secondary SE as well, and our results corroborate this conclusion. Namely, 2,4-D and CPPU successfully induced both primary [13] and secondary SE in *C. erythraea* (Figures 2–5). Similarly, 2,4-D induced both primary and secondary SE in peanut [48] and *Magnolia dealbata* [49], while in some species, hormone-free medium was suitable for efficient induction of both primary and secondary SE [33]. However, there are cases where primary SE is induced by PGRs, but the induction of the secondary SE requires a medium without PGRs for its completion. For example, in carnation, primary ISE was induced through application of 2,4-D and CPPU, while secondary **se** were produced on hormone-free medium [50]. The combination of 2,4-D and CPPU has been investigated during the induction of primary SE [13,29,50–52], but very rarely in the induction of the secondary SE [52–54]. Adventitious (secondary) embryos were formed on 2,4-D and CPPU-containing media in grapevine [53] and *Epipremnum aureum* [52], but this combination of PGRs did not induce secondary **se** in peanut [54].

Another important factor for the efficiency of secondary SE induction is developmental stage of primary embryos. Some protocols for successful secondary SE involved the use of **cse** as explants, for example in *P. menziesii* [34] and *A. trifoliata* [35]. Globular **se** were more suitable for inducing secondary SE than **cse** in some plant species. However, efficient secondary SE was recorded for all developmental stages of somatic embryos (heart, torpedo, cotyledonary) in cabbage and cauliflower [55] and *H. nobilis* [33]. Centaury leaf-derived **cse** exhibited great embryogenic potential in our research, since secondary SE occurred on all tested media (Figure 5b).

The pattern and frequency of secondary embryogenesis, as well as callogenesis, on centaury primary **cse** explants depended on the culture medium composition, where combinations of 2,4-D and CPPU and 2,4-D alone were used to examine the embryogenic capacity of the primary **cse**. Callus formation was observed on all media, but there were obvious differences in the frequency and characteristics of the induced calli among different PGRs combinations and concentrations. Nonembryogenic calli (**nec**) were formed on all types of media (Figure 5a). The highest number of explants forming **nec** only was observed on media supplemented with 2,4-D solely, on media with equal levels of 2,4-D and CPPU, and on media with 0.1 mgL^{-1} 2,4-D and 0.25 mgL^{-1} CPPU (Figure 5a). Further increase of CPPU concentration promoted the formation of **ec** on centaury **cse**, and the highest number of explants with **ec** was obtained on medium with the highest CPPU concentration (Figure 5a). To the best of our knowledge, there are no data concerning the effect of increasing concentrations of CPPU on **ec** formation during secondary SE. During primary SE, the increase in CPPU concentrations resulted in **ec** induction, for example, in carnation [50].

In the absence of PGRs and on the media containing 2,4-D only, somatic embryos were formed only directly (Figures 5b and 6a). Nevertheless, DSE was the predominant developmental pathway even on media containing CPPU (compare "**dcse** only" and "**icse** only" graphs, Figure 5b). Our results are consistent with literature, since in almost all plant species, the origin of embryos in secondary SE is direct, regardless of the PGRs used. Direct secondary SE was induced on hormone-free medium [35], on medium with different cytokinins [56], or on medium with 2,4-D [38]. However, in the present study, along with secondary DSE (Figure 3g,h and Figure 5b), secondary ISE also occurred, since the formation of **ec** (Figures 3c–e and 5a) and **icse** on **cse** (Figure 3f,h and Figure 5b) was observed upon the addition of CPPU in the presence of 2,4-D. Two different pathways—DSE and ISE—on the same **se** of *Castanea sativa* during secondary SE were also observed [57]. The importance of explant type and PGRs in culture medium on the induction of SE in centaury are reviewed in the accompanying article [14]. While in centaury root culture, SE manifests by direct pattern on media without growth regulators [43], SE in leaf culture occurs by indirect pattern and is induced by 2,4-D and CPPU [13]. Current results indicate that the morphogenic response of centaury **cse** is complex and that it could be modulated with different PGRs combinations. For the first time, both DSE and ISE were observed on the same explant in centaury in vitro culture.

Although **se** and **cse** were formed on all media used in this study, the highest number of **cse** per explant (Table 1) and the highest number of explants forming **se** or **cse** (Figure 5b) were obtained on medium with the highest CPPU concentration. This finding is in agreement with Chen and Hong [55], who found that increasing concentrations of a urea-type cytokinin, thidiazuron, significantly enhanced the percentages of secondary SE in *Oncidium* cultivars Gower Ramsey and Sweet Sugar. On the contrary, Szewczyk-Taranek and Pawłowska [33] reported that the highest number of **cse** per explant was detected on medium without PGRs, while the increasing concentrations of cytokinins reduced the frequency of secondary SE. CPPU is a highly active diphenylurea-derived cytokinin with efficiency in the induction of various morphogenetic processes, including SE. Our results showed that the secondary embryo formation occurred at low efficiency on hormone-free medium or medium containing only 2,4-D. In *Tetrapleura tetraptera* [58] and *M. dealbata* [49], secondary **se** were formed on the media supplemented with 2,4-D only.

Histological analysis confirmed that secondary **se** formed both directly (Figure 4d–f) and indirectly (Figure 4g–i) on the primary **cse** explants. Histological examination showed that DSE and ISE originated from the **cse** subepidermal cells (Figure 4d,g), indicating multicellular origin of centaury secondary somatic embryos. Generally, the pathway and the onset of SE are determined by the physiological and morphological characteristics of the plant tissue source from which the explant derived [59]. Induction of DSE is restricted to somatic cells of explants which have acquired embryogenic competence [15]. According to Puigderrajols et al. [60], **se** development in cork oak actually begins when epidermal and subepidermal cell dedifferentiation starts and the entire meristematic proliferation is a PEM.

Hereby, we have provided the structural evidence for the multicellular origin of secondary **se** in *C. erythraea*. We have demonstrated that the secondary **se** of centaury may develop from PEM clusters of rapidly dividing cells of subepidermal layer. Such response may be a consequence of the various degrees of cell maturity of **cse** explants and different content of endogenous hormones [61]. In *C. erythraea*, intensive divisions of the cells in the subepidermal layers of **cse** explants, and the capability of many neighboring cells to act in a coordinated manner, led to the differentiation and efficient embryo development. We speculate that this may be the main reason why secondary somatic embryo formation via the multicellular pathway occurs very quickly. Our results indicate that primary and secondary SE in centaury result from two distinct ontogenetic pathways, DSE and ISE. These two processes led to the production of **cse** and the maintenance of embryogenic competence for more than 3 months. Secondary embryos arose directly from the primary **cse** embryos, where some epidermal and/or subepidermal cells may have already been embryogenically determined [62]. Regarding secondary ISE, the formation of **se** from the **ec** suggested that cotyledon cells divided and proliferated before some of the callus cells had reached embryogenic competence; thus, a callogenesis stage occurred prior to initiation of the embryogenic process.

The process of centaury secondary SE is asynchronous (Figure 3e,h) which is in accordance with secondary SE of other species [32,38]. Asynchronous development of **se** in centaury has been found both during DSE from roots [43] and ISE from leaf explants [13]. Secondary embryos originated from the entire surface of the centaury primary **cse** (Figure 3c,e and Figure 4a–c). In other species, secondary **se** are formed from different parts of primary **se**, for example, from cotyledons [38], hypocotyls [63,64], hypocotyl/root zone [57], roots [65], combinations of these organs, for example, from cotyledons and radicles [66], or stomatal guard cells [56].

The induction of secondary SE in centaury on media containing 0.5 mgL^{-1} CPPU produced **se** with abnormal morphology (Figure 6c,d). Different causes have been proposed for the abnormality in **se** development, such as excessive PGR addition, long exposition, or accumulation of exogenous auxins inside the tissue [67]. Secondary **se** with abnormal morphology could be reused for callus reinduction, as proposed by Ji et al. [37]. Abnormal morphology of secondary embryos could affect their germination and conversion to plantlets. As reported by Ji et al. [37], the best conversion rate had mono- and bi-cotyledonary embryos (about 60%), followed by poly-cotyledonary and trumpet-shaped embryos. Because of this, embryos of abnormal morphology are usually discarded; however, some of them could also be recycled to induce cyclic SE. The response to hormonal treatment can be dependent on the shape of the **cse**, and only fused cotyledonary embryos were successfully induced into **ec** [37]. Maturation and conversion of somatic embryos into plantlets are important processes which enable the establishment of efficient regeneration systems [68]. In the present study, well-developed bi-cotyledonary **cse** were produced predominantly on media containing 0.1 mgL^{-1} 2,4-D and 0.25 mgL^{-1} CPPU and, therefore, these **se** were transferred on media without growth regulators for germination and conversion (Figure 8a–d). Upon transferring, healthy plants were obtained (Figure 8e,f).

Secondary embryogenesis offers the possibility for enhanced production of somatic embryos through establishment of cycling cultures; thus, it is of great importance to determine specific conditions under which cyclic SE occurs [68]. In this study we investigated the effects of two combinations of 2,4-D and CPPU on embryogenic response of **cse** across several cycles (Figures 1 and 7). We found that the embryogenic potential (production of **ec** and **se**) of centaury **cse** explants increased with the cycles' progression. Centaury **cse** showed a high rate of embryogenic callogenesis, which increased with cycles' progression on both media (Figure 7a,b), while the production of **se** increased in the 3rd cycle on lower concentration (Figure 7c), and nearly all explants produced **se** on the higher concentration during two cycles (Figure 7d). Literature encompassing cyclic SE shows that the efficiency of cyclic embryogenesis varies among different species. In *Musa acuminata* AAA cv. Grand Naine, the potential of the explants to produce **se** did not decline with the number of cycles [69]. The embryogenic potential and the mean number of embryos per explant displayed a gradual reduction with subculturing in *A. trifoliata* [35]. Similarly, the percentage of explants with **se** decreased with each cycle of SE induction in *M. dealbata* [49] and in two *Brassica oleracea* varieties [55]. In *C. persicum,* embryogenic competence of calli was affected by number of subculture cycles, since calli from the first cycle showed the highest competence for SE, which decreased during second subculture [32]. Pires et al. [36] developed strategy for recovering and maintaining the cyclic embryogenesis in olive embryogenic calli by its subculturing, which increased the average number of **se** per calli. Therefore, **ec** produced in centaury cyclic embryogenic system could be recycled by transferring onto medium which could enhance maturation of **icse**.

In the present study, the number of explants producing **cse** decreased on both types of media during the cycles (Figure 7c,d). The observed discrepancy in the number of the explants forming **se** of any developmental stage and the number of explants with developed **se** at the cotyledonary stage means that the embryogenic potential of the explants did not decrease, but that the rate of the embryo maturation slowed down with subculturing. Decline in the number of fully-developed **cse** observed in centaury cyclic SE could represent an evidence that long-term maintenance on inductive media affected somatic embryo capacity to advance to later stages. Long exposition to auxin 2.4-D could affect embryo maturation, even though this auxin was proved to be an important factor in the SE induction. Some studies have shown that the presence of 2,4-D in medium is conducive to **ec** induction and proliferation, but that the reduction or removal of 2,4-D promotes **se** development and maturation [38,70]. The **ec** obtained from the immature zygotic embryo of pine trees could produce **se** [71], but the **se** production rate was low, and maturation of **se** was limited. Prolonged culture on the induction medium resulted in an increase in the number of globular and heart-shaped embryos, but did not stimulate the production of mature embryos [64]. In addition, **ec** may lose the potential for SE after extended subculture on medium supplemented with 2,4-D. Although cyclic SE was initiated in both 2,4-D and CPPU combinations in this study, the number of cycles was affected by the higher concentrations of 2,4-D and CPPU. We can speculate that the exposition of **cse**, even to low 2,4-D concentration of 0.1 mgL^{-1}, during cycles disturbed the balance of endogenous phytohormones in the embryogenic explants and delayed the maturation of newly-produced **cse**.

In this study, secondary SE was reported for the first time in centaury. This new morphogenetic response could provide a long-term source of **ec** and **se** by establishing cycling cultures. Complex morphogenic response of centaury **cse** could be modulated with different PGRs combinations. Cyclic SE obtained in this study could be used in centaury as a method for obtaining an amplified pool of SE tissues and especially **cse**, which can germinate into plants with good development characteristics. The developed secondary and cyclic SE system could also have fundamental merit because it allows for biochemical and molecular comparison of **se** obtained from roots, leaves, and primary **cse** explants in centaury.

4. Materials and Methods

4.1. Plant Material

C. erythraea in vitro culture was established as described previously [13]. Commercial seeds (Jelitto Staudensamen GmbH, Schwarmstedt, Germany) were surface-sterilized with bleach (4% hypochlorite) and germinated on $\frac{1}{2}$ MS medium [72] half-strength salts and vitamins, containing 30 gL^{-1} sucrose and solidified with 7 gL^{-1} agar (Torlak, Beograd, Serbia). Seedlings were transferred to the same medium for further growth. All of the cultures were maintained under a 16/8-h (light/dark) photoperiod at irradiance of $47 \text{ µmol m}^{-2} \text{ s}^{-1}$ and temperature of $25 \pm 2 \degree C$.

4.2. Induction of Primary SE

Primary SE was induced according to Filipović et al. [13] with slight modifications. Leaf segments were dissected from well-developed, two-month-old in vitro grown plants and cultured, abaxial side down, in Petri dishes containing basal medium formulation consisting of MS salts and vitamins and supplemented with 30 gL^{-1} sucrose, 7 gL^{-1} agar, 0.1 mgL^{-1} 2,4-D (Sigma-Aldrich, Steinheim, Germany) and 0.25 mgL^{-1} CPPU (Sigma-Aldrich, Germany). The leaf explants were maintained in darkness, at temperature of 25 ± 2 °C, and **se** were developed during three weeks. The obtained primary **se** at the cotyledonary stage (**cse**) were used as explants for the induction of secondary and cyclic SE. In parallel, primary SE was also induced on higher PGRs concentrations of 0.2 mgL^{-1} 2,4-D and 0.5 mgL^{-1} CPPU, and the **cse** obtained on this concentration were used in the experiment with the induction of cyclic SE (Figure 1).

4.3. Induction of Secondary SE

Primary **cse** obtained on the medium with 0.1 mgL^{-1} 2,4-D and 0.25 mgL^{-1} CPPU were transferred to five different media: a hormone-free MS medium, and media containing 0.1 mgL^{-1} 2,4-D and CPPU at increasing concentrations (0, 0.1, 0.25, and 0.5 mgL^{-1}, see Figure 1). The cultures were kept in constant darkness at 25 ± 2 °C. The experiment was performed in seven replicates (Petri dishes) per treatment, with 36 **cse** per replicate. The **cse** explants were systematically arranged in 6×6 arrays, numbered, and documented both photographically and by observing under a binocular microscope (Leica WILD, MPS 28/32, M3Z, Wetzlar, Germany). Developmental parameters, such as the number of explants forming calli or secondary embryos and the number of secondary **cse** developed per explant, were recorded after four weeks.

4.4. Induction of Cyclic SE

For the induction of cyclic SE, primary **cse**, obtained from leaf explants on medium containing lower (0.1 mgL^{-1} 2,4-D and 0.25 mgL^{-1} CPPU) or higher concentration of PGRs (0.2 mgL^{-1} 2,4-D and 0.5 mgL^{-1} CPPU), were dissected and transferred to the same medium. The cultures were kept in constant darkness at 25 ± 2 °C. In the 1st cycle, secondary **cse** that developed on either of the two media were transferred to fresh media maintaining the same treatment (Figure 1). In the 2nd cycle, tertiary **cse** that formed after four weeks were subcultured on the same fresh media for the 3rd and last cycle. For both lower and higher hormone treatments, the experiment was performed in four replicates with 36 **cse** explants per Petri dish. After four weeks in each cycle, developmental parameters (as in Section 4.3) were recorded.

4.5. Germination of Somatic Embryos

The viability of secondary **cse** obtained on medium containing 0.1 mgL^{-1} 2,4-D and 0.25 mgL^{-1} CPPU was evaluated in terms of somatic embryo germination and plantlet conversion. White opaque secondary **cse**, 1.5–2.5 mm in length, were excised from the primary **cse** explants and set to germinate on MS media without PGRs. The germination experiment was conducted in the light with 105 secondary **cse** and in darkness with 75 **cse**. The germination was scored after 25 days in culture as the percentage of **cse** developing shoots and/or roots.

4.6. Histological Analysis

Ontogeny of secondary somatic embryos obtained on medium with 0.1 mgL^{-1} 2,4-D and 0.25 mgL^{-1} CPPU was studied by histological analysis. To confirm histologically that secondary SE was indeed induced in the dark, primary **cse** explants obtained on this medium were transferred to the same medium, kept in the dark, and sampled after three weeks. For histological analysis, the **cse** explants, along with developing **se**, were fixed in mixture of formalin–glacial acetic acid–70% ethanol (FAA) for 24 h, dehydrated in an ethanol series, and embedded in Histowax (Histolab, Västra Frölunda, Sweden) at 56–58 °C.

Five-μm-thick sections were cut using a Reichter rotary microtome (Reichter, Wien, Austria) and stained with haematoxylin [73]. The sections were observed and photographed under appropriate magnifications using Nikon Eclipse E100 light research microscope (Nikon, Tokyo, Japan). All images were recorded with Bresser MikroCam SP 5.1 camera and software (Bresser, Rhede, Germany).

4.7. Data Collecting and Statistical Analysis

The processes of secondary SE on five different media, as well as cyclic SE on two different media, through the three cycles were evaluated by scoring several developmental parameters (events) after four weeks in culture. The scored developmental parameters included the number of **cse** explants that developed: any type of calli, embryogenic calli (**ec**) or exclusively nonembryogenic calli (**nec**), as well as the number of **cse** explants that developed: secondary somatic embryos at any stage or origin (**se**), secondary somatic embryos at the cotyledonary stage (**cse**), and specifically **cse** formed by direct (**dcse**) or indirect path (**icse**). All statistical analyses were performed using the R programming language for statistical computing [74].

The effect of PGRs concentration on the occurrence of specific developmental events on the explants was analyzed with logistic regression using quasi-binomial distribution to account for overdispersion and logit link function. Medium formulations where no explants induced specific types of calli or embryos were not included in the logistic regression models. The statistical significance of the effect of medium formulation was evaluated using likelihood ratio tests, and for parameters where the effect was significant ($p < 0.05$), pairwise comparisons were performed using the emmeans R package [75]. To account for multiple comparisons, Bonferroni correction was applied, and adjusted p-values < 0.05 were considered statistically significant. Bar height on graphs represents fitted values for logistic regression models and error bars represent 95% confidence intervals. Statistically significant differences are denoted with a compact letter display on the figures. Average number of secondary **cse** per experimental replicate was analyzed using ANOVA with Tukey post hoc test for pairwise comparisons.

Data from the cyclic SE experiment were analyzed using generalized estimating equations, as implemented in R package geepack [76]. This approach was chosen since the observations between the cycles were not independent: **cse** obtained in the replicate (Petri dish) no. 1 of the 1st cycle were used as explants in the replicate no. 1 of the 2nd cycle etc., maintaining the same treatment. Occurrence of specific types of calli or embryos on the explants was analyzed using logistic generalized estimating equations, while average number of secondary **cse** per experimental replicate was analyzed using generalized estimating equations. The statistical significance of the effect of SE cycle was evaluated by Wald test statistic [76], and for parameters where the effect was significant ($p < 0.05$), pairwise comparisons were performed using the emmeans R package [75]. To account for multiple comparisons, Bonferroni correction was applied, and adjusted p-values < 0.05 were considered statistically significant. Bar height on graphs represents fitted values for general estimation equations (GEE) logistic regression models and error bars represent 95% confidence intervals. Statistically significant differences are denoted with a compact letter display in figures.

Author Contributions: Conceptualization, experimental design, and supervision: M.D.B. and S.I.T.; experimental work: S.I.T., M.D.B., K.B.Ć., and A.R.S.; data interpretation: S.I.T., M.D.B., M.B.D. and A.R.S.; writing—original draft preparation: A.R.S., M.D.B. and B.K.F.; statistical analysis: M.B.D.; writing—review and editing: A.D.S.; visualization: M.D.B., A.R.S., S.I.T. and A.D.S. All authors have read and agreed to the published version of the manuscript

Funding: This work was funded by the Ministry of Education, Science and Technological Development of the Republic of Serbia, Contract 451-03-68/2020-14/200007.

Data Availability Statement: Representations of the data are contained in the article, raw data is available upon request.

Conflicts of Interest: The authors declare no conflict of interest.

References

1. Schouppe, D.; Rein, B.; Vallejo-Marin, M.; Jacquemyn, H. Geographic variation in floral traits and the capacity of autonomous selfing across allopatric and sympatric populations of two closely related *Centaurium* species. *Sci. Rep.* **2017**, *7*, 46410. [CrossRef]
2. Schat, H.; Ouborg, J.; de Wit, R. Life history and plant architecture: Size-dependent reproductive allocation in annual and biennial *Centaurium* species. *Acta Bot. Neerl.* **1989**, *38*, 183–201. [CrossRef]
3. Tuluce, Y.; Ozkol, H.; Koyuncu, I.; Ine, H. Gastroprotective effect of small centaury (*Centaurium erythraea* L.) on aspirin-induced gastric damage in rats. *Toxicol. Ind. Health* **2011**, *27*, 760–768. [CrossRef] [PubMed]
4. Mroueh, M.; Saab, Y.; Rizkallah, R. Hepatoprotective activity of *Centaurium erythraea* on acetaminophen-induced hepatotoxicity in rats. *Phytother. Res.* **2004**, *18*, 431–433. [CrossRef] [PubMed]
5. Valentao, P.; Andrade, P.B.; Silva, E.; Vicente, A.; Santos, H.; Bastos, M.L.; Seabra, R.M. Methoxylated xanthones in the quality control of small centaury *(Centaurium erythraea)* flowering tops. *J. Agr. Food Chem.* **2002**, *50*, 460–463. [CrossRef]
6. Subotič, A.; Jankovič, T.; Jevremovič, S.; Grubišič, D. Plant Tissue Culture and Secondary Metabolites Productions of *Centaurium erythraea* Rafn., a Medical plant. In *Floriculture, Ornamental and Plant Biotechnology: Advances and Topical Issues*, 1st ed.; Teixeira da Silva, J.A., Ed.; Global Science Books: London, UK, 2006; Volume 2, pp. 564–570.
7. Šiler, B.; Živković, S.; Banjanac, T.; Cvetković, J.; Nestorović-Živković, J.; Ćirić, A.; Soković, M.; Mišić, D. Centauries as underestimated food additives: Antioxidant and antimicrobial potential. *Food Chem.* **2014**, *147*, 367–376. [CrossRef]
8. Jovanović, O.; Radulović, N.; Stojanović, G.; Palić, R.; Zlatković, B.; Gudžić, B. Chemical composition of the essential oil of *Centaurium erythraea* Rafn (Gentianaceae) from Serbia. *J. Essent. Oil Res.* **2009**, *21*, 317–322. [CrossRef]
9. Kirbag, S.; Zengin, F.; Kursat, M. Antimicrobial activities of extracts of some plants. *Pak. J. Bot.* **2009**, *41*, 2067–2070.
10. Đorđević, M.; Grdović, N.; Mihailović, M.; Arambašić-Jovanović, J.; Uskoković, A.; Rajić, J.; Đordjević, M.; Tolić, A.; Mišić, D.; Šiler, B.; et al. *Centaurium erythraea* methanol extract protects red blood cells from oxidative damage in streptozotocin-induced diabetic rats. *J. Ethnopharmacol.* **2017**, *202*, 172–183. [CrossRef]
11. Hamza, N.; Berke, B.; Cheze, C.; Agli, A.N.; Robinson, P.; Gin, H.; Moore, N. Prevention of type 2 diabetes induced by high fat diet in the C57BL/6J mouse by two medicinal plants used in traditional treatment of diabetes in the east of Algeria. *J. Ethnopharmacol.* **2010**, *128*, 513–518. [CrossRef]
12. Đorđević, M.; Grdović, N.; Mihailović, M.; Jovanović, J.A.; Uskoković, A.; Rajić, J.; Sinadinović, M.; Tolić, A.; Mišić, D.; Šiler, B.; et al. *Centaurium erythraea* extract improves survival and functionality of pancreatic beta-cells in diabetes through multiple routes of action. *J. Ethnopharmacol.* **2019**, *242*, 112043. [CrossRef] [PubMed]
13. Filipović, B.K.; Simonović, A.D.; Trifunović, M.M.; Dmitrović, S.S.; Savić, J.M.; Jevremović, S.B.; Subotić, A.R. Plant regeneration in leaf culture of *Centaurium erythraea* Rafn. Part 1: The role of antioxidant enzymes. *Plant Cell Tissue Organ. Cult.* **2015**, *121*, 703–719. [CrossRef]
14. Simonović, A.D.; Trifunović-Momčilov, M.M.; Filipović, B.K.; Marković, M.P.; Bogdanović, M.D.; Subotić, A.R. Somatic Embryogenesis in *Centaurium erythraea* Rafn—Current Status and Perspectives: A Review. *Plants* **2021**, *10*, 70. [CrossRef] [PubMed]
15. Quiroz-Figueroa, F.R.; Rafael, R.H.; Galaz-Avalos, R.M.; Loyola-Vargas, V.M. Embryo production through somatic embryogenesis can be used to study cell differentiation in plants. *Plant Cell Tissue Organ. Cult.* **2006**, *86*, 258–301. [CrossRef]
16. Karami, O.; Aghavaisi, B.; Pour, A.M. Molecular aspects of somatic-to-embryogenic transition in plants. *J. Chem. Biol.* **2009**, *2*, 177–190. [CrossRef]
17. Fehér, A. Somatic embryogenesis—Stress-induced remodeling of plant cell fate. *Biochim. Biophys. Acta* **2015**, *1849*, 385–402. [CrossRef]
18. Raemakers, C.J.J.M.; Jacobsen, E.; Visser, R.G.F. Secondary somatic embryogenesis and applications in plant breeding. *Euphytica* **1995**, *81*, 93–107. [CrossRef]
19. Guan, Y.; Li, S.G.; Fan, X.F.; Su, Z.H. Application of somatic embryogenesis in woody plants. *Front. Plant Sci.* **2016**, *7*, 938. [CrossRef]
20. Giri, C.C.; Shyamkumar, B.; Anjaneyulu, C. Progress in tissue culture, genetic transformation and applications of biotechnology to trees: An overview. *Trees* **2004**, *18*, 115–135. [CrossRef]
21. Engelmann, F. Use of biotechnologies for the conservation of plant biodiversity. *Vitro Cell Dev. Biol. Plant* **2011**, *47*, 5–16. [CrossRef]
22. Ibáñez, S.; Carneros, E.; Testillano, P.S.; Pérez-Pérez, J.M. Advances in Plant Regeneration: Shake, Rattle and Roll. *Plants* **2020**, *9*, 897. [CrossRef] [PubMed]
23. Altamura, M.; Della Rovere, F.; Fattorini, L.; D'Angeli, S.; Falasca, G. Recent Advances on Genetic and Physiological Bases of *In Vitro* Somatic Embryo Formation. In *In Vitro Embryogenesis in Higher Plants*, 1st ed.; Germana, M., Lambardi, M., Eds.; Humana Press-Springer: New York, NY, USA, 2016; Volume 1359, pp. 47–85. [CrossRef]
24. Loyola-Vargas, V.; Ochoa-Alejo, N. Somatic embryogenesis. An overview. In *Somatic Embryogenesis: Fundamental Aspects and Applications*, 1st ed.; Loyola-Vargas, V., Ochoa-Alejo, N., Eds.; Springer International Publishing: Cham, Switzerland, 2016; pp. 1–8. [CrossRef]
25. Tomiczak, K.; Mikuła, A.; Niedziela, A.; Wójcik-Lewandowska, A.; Domżalska, L.; Rybczyński, J.J. Somatic embryogenesis in the family Gentianaceae and its biotechnological application. *Front. Plant Sci.* **2019**, *10*, 762. [CrossRef] [PubMed]
26. Fehér, A. The initiation phase of somatic embryogenesis: What we know and what we don't. *Acta Biol. Szeged.* **2008**, *52*, 53–56.

27. Jiménez, V.M.; Bangerth, F. Endogenous hormone levels in explants and in embryogenic and non-embryogenic cultures of carrot. *Phys. Plant* **2001**, *111*, 389–395. [CrossRef] [PubMed]
28. Vondráková, Z.; Krajňáková, J.; Fischerová, L.; Vágner, M.; Eliášová, K. Physiology and role of plant growth regulators in somatic embryogenesis. In *Vegetative Propagation of Forest Trees*; Park, Y.S., Bonga, J.M., Moon, H.K., Eds.; National Institute of Forest Science: Seoul, Korea, 2016; pp. 123–169.
29. Fiuk, A.; Rybczynski, J.J. Genotype and plant growth regulator dependent response of somatic embryogenesis from *Gentiana* spp. leaf explants. *Vitro Cell Dev. Biol. Plant* **2008**, *44*, 90–99. [CrossRef]
30. Cantelmo, L.; Soares, B.O.; Rocha, L.P.; Pettinelli, J.A.; Callado, C.H.; Mansur, E.; Casteller, A.; Gagliardi, R.F. Repetitive somatic embryogenesis from leaves of the medicinal plant *Petiveria alliacea* L. *Plant Cell Tissue Organ. Cult.* **2013**, *115*, 385–393. [CrossRef]
31. Yang, X.; Lu, J.; Teixeira da Silva, J.M.; Ma, G. Somatic embryogenesis and shoot organogenesis from leaf explants of *Primulina tabacum*. *Plant Cell Tissue Organ. Cult.* **2012**, *109*, 213–221. [CrossRef]
32. You, C.R.; Fan, T.J.; Gong, X.Q.; Bian, F.H.; Liang, L.K.; Qu, F.N. A high-frequency cyclic secondary somatic embryogenesis systemfor *Cyclamen persicum* Mill. *Plant Cell Tissue Organ. Cult.* **2011**, *107*, 233–242. [CrossRef]
33. Szewczyk-Taranek, B.; Pawłowska, B. Recurrent somatic embryogenesis and plant regeneration from seedlings of *Hepatica nobilis* Schreb. *Plant Cell Tissue Organ. Cult.* **2015**, *120*, 1203–1207. [CrossRef]
34. Lelu-Walter, M.A.; Gautier, F.; Eliášová, K.; Sanchez, L.; Teyssier, C.; Lomenech, A.M.; Le Mette, C.; Hargreaves, C.; Trontin, J.F.; Reeves, C. High gellan gum concentration and secondary somatic embryogenesis: Two key factors to improve somatic embryo development in *Pseudotsuga menziesii* [Mirb.]. *Plant Cell Tissue Organ. Cult.* **2018**, *132*, 137–155. [CrossRef]
35. Zou, S.; Yao, X.; Zhong, C.; Shuaiyu, Z.; Xiaohong, Y.; Caihong, Z.; Dawei, L.; Zupeng, W.; Hongwen, H. Recurrent somatic embryogenesis and development of somatic embryos in *Akebia trifoliata* (Thunb.) Koidz (Lardizabalaceae). *Plant Cell Tiss Organ. Cult.* **2019**, *139*, 493–504. [CrossRef]
36. Pires, R.; Cardoso, H.; Ribeiro, A.; Peixe, A.; Cordeiro, A. Somatic embryogenesis from mature embryos of *Olea europaea* L. cv. 'Galega Vulgar' and long-term management of calli morphogenic capacity. *Plants* **2020**, *9*, 758. [CrossRef] [PubMed]
37. Ji, W.; Luo, Y.; Guo, R.; Li, X.; Zhou, Q.; Ma, X.; Wang, Y. Abnormal Somatic Embryo Reduction and Recycling in Grapevine Regeneration. *J. Plant Growth Regul.* **2017**, *36*, 912–918. [CrossRef]
38. Karami, O.; Deljou, A.; Kordestani, G.K. Secondary somatic embryogenesis of carnation (*Dianthus caryophyllus* L.). *Plant Cell Tiss. Organ. Cult.* **2008**, *92*, 273–280. [CrossRef]
39. Klimaszewska, K.; Noceda, C.; Pelletier, G.; Label, P.; Rodriguez, R.; Lelu-Walter, M.-A. Biological characterization of young and aged embryogenic cultures of *Pinus pinaster* (Ait.). *In Vitro Cell Dev. Biol. Plant* **2009**, *45*, 20–33. [CrossRef]
40. Pila Quinga, L.A.; Pacheco de Freitas Fraga, H.; do Nascimento Vieira, L.; Guerra, M.P. Epigenetics of long-term somatic embryogenesis in *Theobroma cacao* L.: DNA methylation and recovery of embryogenic potential. *Plant Cell Tissue Organ. Cult.* **2017**, *131*, 295–305. [CrossRef]
41. Barešová, H.; Kamínek, M. Light induce embryogenesis in suspension culture of *Centaurium erythraea*. In *International Symposium Plant Tissue and Cell Culture Application to Crop Improvement*; Novák, F.J., Havel, L., Doležel, J., Eds.; Czechoslovak Academy of Sciences: Prague, Czechoslovakia, 1984; pp. 163–164.
42. Subotić, A.; Budimir, S.; Grubišić, D.; Momčilović, I. Direct regeneration of shoots from hairy root cultures of *Centaurium erythraea* inoculated with *Agrobacterium rhizogenes*. *Biol. Plantarum.* **2003**, *47*, 617–619. [CrossRef]
43. Subotić, A.; Grubišić, D. Histological analysis of somatic embryogenesis and adventitious formation from root explants of *Centaurium erythreae* Gillib. *Biol. Plant* **2007**, *51*, 514–516. [CrossRef]
44. Subotić, A.; Jevremović, S.; Grubišić, D. Influence of cytokinins on *in vitro* morphogenesis in root cultures of *Centaurium erythraea*—valuable medicinal plant. *Sci. Hortic.* **2009**, *120*, 386–390. [CrossRef]
45. Subotić, A.; Jevremović, S.; Grubišić, D.; Janković, T. Spontaneous plant regeneration and production of secondary metabolites from hairy root cultures of *Centaurium erythraea* Rafn. In *Protocols for In Vitro Cultures and Secondary Metabolite Analysis of Aromatic and Medicinal Plants, Methods in Molecular Biology*; Jain, S.M., Saxena, P.K., Eds.; Springer: Berlin/Heidelberg, Germany, 2009; Volume 547, pp. 205–215. [CrossRef]
46. Simonović, A.D.; Filipović, B.K.; Trifunović, M.M.; Malkov, S.N.; Milinković, V.P.; Jevremović, S.B.; Subotić, A.R. Plant regeneration in leaf culture of *Centaurium erythraea* Rafn. Part 2: The role of arabinogalactan proteins. *Plant Cell Tissue Organ. Cult.* **2015**, *121*, 721–739. [CrossRef]
47. Ćuković, K.; Dragićević, M.; Bogdanović, M.; Paunović, D.; Giurato, G.; Filipović, B.; Subotić, A.; Todorović, S.; Simonović, A. Plant regeneration in leaf culture of *Centaurium erythraea* Rafn. Part 3: De *novo* transcriptome assembly and validation of housekeeping genes for studies of *in vitro* morphogenesis. *Plant Cell Tissue Organ. Cult.* **2020**, 1–17. [CrossRef]
48. Baker, C.M.; Wetzstein, H.Y. Repetitive somatic embryogenesis in peanut cotyledon cultures by continual exposure to 2,4-D. *Plant Cell Tissue Organ. Cult.* **1995**, *40*, 249–254. [CrossRef]
49. Chávez-Cortazar, A.; Mata-Rosas, M.; Oyama, K.; Samain, M.S.; Quesada, M. Induction of somatic embryogenesis and evaluation of genetic stability in regenerated plants of *Magnolia dealbata*. *Biol. Plant* **2020**, *64*, 224–233. [CrossRef]
50. Aalifar, M.; Arab, M.; Aliniaeifard, S.; Dianati, S.; Mehrjerdi, M.Z.; Limpens, E.; Serek, M. Embryogenesis efficiency and genetic stability of *Dianthus caryophyllus* embryos in response to different light spectra and plant growth regulators. *Plant Cell Tissue Organ. Cult.* **2019**, *139*, 479–492. [CrossRef]

51. Fiore, S.; De Pasquale, F.; Carimi, F.; Sajeva, M. Effect of 2,4-D and 4-CPPU on somatic embryogenesis from stigma and style transverse thin cell layers of Citrus. *Plant Cell Tissue Organ. Cult.* **2002**, *68*, 57–63. [CrossRef]

52. Zhang, Q.; Chen, J.; Henny, R.J. Direct somatic embryogenesis and plant regeneration from leaf, petiole, and stem explants of Golden Pothos. *Plant Cell Rep.* **2005**, *23*, 587–595. [CrossRef]

53. Nakano, M.; Sakakibara, T.; Watanabe, Y.; Mii, M. Establishment of embryogenic cultures in several cultivars of *Vitis vinifera* and *V. x labruscana*. *Vitis* **1997**, *36*, 141–145.

54. Little, E.L.; Magbanua, Z.V.; Parrott, W.A. A protocol for repetitive somatic embryogenesis from mature peanut epicotyls. *Plant Cell Rep.* **2000**, *19*, 351–357. [CrossRef]

55. Pavlović, S.; Vinterhalter, B.; Zdravković-Korać, S.; Zdravković, J.; Cvikić, D.; Mitić, N. Recurrent somatic embryogenesis and plant regeneration from immature zygotic embryos of cabbage (*Brassica oleracea* var. capitata) and cauliflower (*Brassica oleracea* var. botrytis). *Plant Cell Tissue Organ. Cult.* **2013**, *113*, 397–406. [CrossRef]

56. Chen, J.T.; Hong, P.I. Cellular origin and development of secondary somatic embryos in *Oncidium* leaf cultures. *Biol. Plant* **2012**, *56*, 215–220. [CrossRef]

57. Corredoira, E.; Ballester, A.; Vieitez, A.M. Proliferation, Maturation and Germination of Castanea sativa Mill. Somatic Embryos Originated from Leaf Explants. *Ann. Bot* **2003**, *92*, 129–136. [CrossRef] [PubMed]

58. Opabode, J.T.; Akinyemiju, O.A.; Ayeni, O.O. Plant Regeneration via Somatic Embryogenesis from Immature Leaves in *Tetrapleura tetraptera* (Schum. & Thonn.) Taub. *Arch. Biol. Sci.* **2011**, *63*, 1135–1145. [CrossRef]

59. Gaj, M.D. Factors influencing somatic embryogenesis induction and plant regeneration with particular reference to *Arabidopsis thaliana* (L.) Heynh. *Plant Growth Regul.* **2004**, *43*, 27–47. [CrossRef]

60. Puigderrajols, P.; Mir, G.; Molinas, M. Ultrastructure of early secondary embryogenesis by multicellular and unicellular pathways in cork oak (*Quercus suber* L.). *Ann. Bot* **2001**, *87*, 179–189. [CrossRef]

61. Grzyb, M.; Kalandyk, A.; Mikuła, A. Effect of TIBA, fluridone and salicylic acid on somatic embryogenesis and endogenous hormone and sugar contents in the tree fern *Cyathea delgadii* Sternb. *Acta Physiol. Plant* **2018**, *40*, 1. [CrossRef]

62. Williams, E.; Maheswaran, G.G. Somatic Embryogenesis: Factors Influencing Coordinated Behaviour of Cells as an Embryogenic Group. *Ann. Bot.* **1986**, *57*, 443–462. [CrossRef]

63. Fernández-Da Silva, R.; Hermoso-Gallardo, L.; Menéndez-Yuffá, A. Primary and secondary somatic embryogenesis in leaf sections and cell suspensions of *Coffea arabica* cv. Catimor. *Interciencia* **2005**, *30*, 694–698. Available online: https://www.redalyc.org/articulo.oa?id=33911006 (accessed on 16 January 2021).

64. Chitra Devi, B.; Narmathabai, V. Somatic embryogenesis in the medicinal legume *Desmodium motorium* (Houtt.) Merr. *Plant Cell Tiss Organ. Cult.* **2011**, *106*, 409–418. [CrossRef]

65. Nair, R.R.; Dutta Gupta, S. High-frequency plant regeneration through cyclic secondary somatic embryogenesis in black pepper (*Piper nigrum* L.). *Plant Cell Rep.* **2006**, *24*, 699–707. [CrossRef]

66. Ćalić, D.; Zdravković-Korać, S.; Radojević, L. Secondary embryogenesis in androgenic embryo cultures of *Aesculus hippocastanum* L. *Biol. Plant* **2005**, *49*, 435–438. [CrossRef]

67. Garcia, C.; de Almeida, A.A.F.; Costa, M.; Britto, D.; Valle, R.; Royaert, S.; Marelli, J.P. Abnormalities in somatic embryogenesis caused by 2,4-D: An overview. *Plant Cell Tiss Organ. Cult.* **2019**, *137*, 193–212. [CrossRef]

68. Mazri, M.A.; Naciri, R.; Belkoura, I. Maturation and Conversion of Somatic Embryos Derived from Seeds of Olive (*Olea europaea* L.) cv. Dahbia: Occurrence of Secondary Embryogenesis and Adventitious Bud Formation. *Plants* **2020**, *9*, 1489. [CrossRef] [PubMed]

69. Remakanthan, A.; Menon, T.G.; Soniya, E.V. Somatic embryogenesis in banana (*Musa acuminata* AAA cv. Grand Naine): Effect of explant and culture conditions. *Vitro Cell Dev. Biol. Plant* **2014**, *50*, 127–136. [CrossRef]

70. Ali, M.; Mujib, A.; Tonk, D.; Zafar, N. Plant regeneration through somatic embryogenesis and genome size analysis of *Coriandrum sativum* L. *Protoplasma* **2017**, *254*, 343–352. [CrossRef] [PubMed]

71. Gao, F.; Pen, C.X.; Wang, H.; Shen, H.L.; Yand, L. Selection of culture conditions for callus induction and proliferation by somatic embryogenesis of *Pinus koraiensis*. *J. For. Res.* **2020**. [CrossRef]

72. Murashige, T.; Skoog, F. A revised medium for rapid growth and bioassays with tobacco tissue cultures. *Physiol. Plant* **1962**, *15*, 473–479. [CrossRef]

73. Johansen, D.A. *Plant Microtechnique*; McGraw-Hill Book Co.: New York, NY, USA, 1940; Volume 3.

74. R Core Team. *R: A Language and Environment for Statistical Computing*; R Foundation for Statistical Computing: Vienna, Austria, 2020; Available online: https://www.R-project.org/Version4.02 (accessed on 16 January 2021).

75. Lenth, R. Emmeans: Estimated Marginal Means, aka Least-Squares Means. R Package Version 1.5.2-1. 2020. Available online: https://CRAN.R-project.org/package=emmeans (accessed on 16 January 2021).

76. Halekoh, U.; Højsgaard, S.; Yan, J. The R Package Geepack for Generalized Estimating Equations. *J. Stat. Softw.* **2006**, *15*, 1–11. Available online: http://hdl.handle.net/10.18637/jss.v015.i02 (accessed on 16 January 2021). [CrossRef]

Article

Early Low-Fluence Red Light or Darkness Modulates the Shoot Regeneration Capacity of Excised *Arabidopsis* Roots

Xi Wei [1,2,†], Yanpeng Ding [2,†], Ye Wang [2], Fuguang Li [1,2,*] and Xiaoyang Ge [2,*]

[1] Henan Normal University Research Base of State Key Laboratory of Cotton Biology, Xinxiang 453000, China; chinaweixi521@163.com
[2] State Key Laboratory of Cotton Biology, Institute of Cotton Research of CAAS, Anyang 455000, China; 15369229109@163.com (Y.D.); wangye01@caas.cn (Y.W.)
* Correspondence: aylifug@caas.cn (F.L.); gexiaoyang@caas.cn (X.G.)
† These authors are contributed equally.

Received: 8 September 2020; Accepted: 13 October 2020; Published: 16 October 2020

Abstract: In plants, light is an important environmental signal that induces meristem development and interacts with endogenous signals, including hormones. We found that treatment with 24 h of low-fluence red light (24 h R) or 24 h of darkness (24 h D) following root excision greatly increased the frequency of shoot generation, while continuous low-fluence red light in callus and shoot induction stages blocked the explants' ability to generate shoots. Shoot generation ability was closely associated with *WUS* expression and distribution pattern. 1-N-naphthylphtalamic acid (NPA) disrupted the dynamic distribution of the *WUS* signal induced by early 24 h R treatment, and NPA plus 24 R treatment increased the average shoot number compared with early 24 h R alone. Transcriptome analysis revealed that differentially expressed genes involved in meristem development and hormone signal pathways were significantly enriched during 24 R or 24 D induced shoot regeneration, where early 24 h R or 24 h D treatment upregulated expression of *WOX5*, *LBD*16, *LBD*18 and *PLT*3 to promote callus initiation and formation of root primordia, and also activated *WUS*, *STM*, *CUC*1 and *CUC*2 expression, leading to initiation of the shoot apical meristem (SAM). This finding demonstrates that early exposure of explants to transient low-fluence red light or darkness modulates the expression of marker genes related with callus development and shoot regeneration, and dynamic distribution of *WUS*, leading to an increased ability to generate shoots.

Keywords: low-fluence; red light; shoot regeneration; *WUS*; NPA

---

1. Introduction

Plant cells are pluripotent, meaning they have the potential to develop into an entire plant body from highly differentiated tissues or organs, or from a single somatic cell [1]. Explants have the ability to regenerate new root apical meristems (RAM) or shoot apical meristems (SAM) in the absence of sexual fertilization [2,3]. Regeneration in differentiation processes can be divided into two categories, including somatic embryogenesis and somatic organogenesis. Somatic organogenesis is important for transgenic plant generation [4,5]; shoot regeneration can be induced from callus tissues culture in two phases. In the first phase, explants of excised *Arabidopsis* root or cotyledon are cultured on a callus-induction medium (CIM) under dark conditions to induce callus formation [5]. Callus cells form when the plant tissue becomes dedifferentiated and acquires pluripotency, which is necessary for shoot regeneration [6,7]. Some studies have shown that callus initiation on a CIM is similar to the rooting pathway in non-root organs where the newly formed callus resembles a group of root primordium-like cells [8,9]. Auxin and cytokinin, involved in somatic organogenesis, may exhibit similar function to that in lateral root development, wherein auxin triggers lateral root initiation but cytokinin inhibits lateral root formation [10]. Furthermore, acquisition of pluripotency in callus cells is also regulated by PLETHORA3 (*PLT3*), *PLT5*, and *PLT7* genes [11]. In the second stage, the callus is transferred to a shoot-induction medium (SIM) to induce the shoots. The shoot induction process consists of several critical events including the distribution of phytohormones over a gradient, initiation of the shoot meristem, and organ formation [1,12]. The CUP-SHAPED COTYLEDON 1 (*CUC1*) and *CUC2*, as No apical meristem, Arabidopsis transcription activation factor, Cup-shaped cotyledon(NAC) transcription factors, regulate the initiation of shoot meristem tissue and promote adventitious shoot regeneration by activating expression of *STM* [13]. A *cuc1 cuc2* double knockout mutation impairs the capacity for shoot regeneration in the callus, while overexpression of *CUC1* or *CUC2* improves the capacity for shoot regeneration [11]. A specific ratio of auxin and cytokinin is key for ensuring *WUS* induction at an appropriate expression level during de novo shoot regeneration in *Arabidopsis* [1,12,14]. *WUS* expression is activated by cytokinin response regulatory factors B-type *ARABIDOPSIS* RESPONSE REGULATORs (*ARRs*) in regions with high levels of cytokinin, which leads to a cell fate transition from callus pluripotency cells to stem cells [15,16].

Light, a critical environmental signal, also modulates shoot regeneration, and has profound developmental effects on shoot organogenesis [17]. After tissue excision, low or high intensity light treatment can affect shoot regeneration in multiple plant species [18]. *Arabidopsis* explants are typically placed in continuous darkness or white light immediately after excision [6]. As for the effects of treatment with specific colors of light, some studies revealed that shoot regeneration was inhibited by blue/UV-A wavelengths, since high-energy wavelengths are absorbed by chlorophyll, thus leading to photosystem II damage [19]. Blue/UV-A wavelengths, even in low fluence light, can inhibit long-term shoot regeneration via a *CRY1* photoreceptor-mediated signaling pathway [20]. High intensity light reduces the ability for shoot regeneration in *Arabidopsis* explants in vitro. Previous studies found that light affects multiple signaling pathways involving auxin [21], cytokinin [22], ethylene [23], red/far-red light photoactivation [24,25], blue/UV-A light photoactivation [26], and photo-oxidative damage [18]. Shoot regeneration was inhibited by treatment with 24 h blue/UV-A wavelengths after organ excision, while far red light signaling counteracts the inhibitory effects on shoot regeneration of early high intensity light exposure [20]. However, it is still not clear what mechanisms underly light regulation of adventitious shoot meristem formation as well as the role of early red light signaling on modulating the efficiency of shoot regeneration.

Red light, a component of sunlight, is of great importance for plant development. Exposure to red light significantly effects morphology, enzymatic activities, and the accumulation of bioactive compounds in *Anoectochilus roxburghii* [27]. The appropriate combination of red and blue wavelengths during embryogenic callus differentiation promotes somatic embryo maturation and conversion in sugarcane [28]. Low flux red light enhances the synthesis of endogenous auxin in *Arabidopsis* meristems [29,30]. We inferred from these reports that low flux red light may also play an important role during shoot organogenesis. In this study, we found that exposure of explants to long-term low-fluence red light strongly inhibited the generation of adventitious shoots, while 24 h exposure to low-fluence red light after root excision significantly improved the efficiency of shoot regeneration.

2. Results

2.1. Effects of Different Light Combinations on Shoot Regeneration Capacity

Root explants from wild-type Arabidopsis Col-0 were used to evaluate the effects of different combinations of light on the capacity for shoot regeneration, through treatments applied during the CIM and SIM stages (Figure 1A). The shoot regeneration capacity was calculated at 10 days, 14 days and 16 days on SIM following the different treatments, respectively. The conditions of the control culture for root explants were first callus induction under seven days of darkness on a CIM, followed by shoot generation on a SIM under white light (D-W). Compared with the control D-W, continuous red light treatment for 7 days on a CIM and white light treatment on a SIM (R-W) caused no significant differences in shoot formation at 10 days and 14 days of shoot induction, but showed an obviously decreased capacity for shoot generation at 16 days, where the percent of explants with shoots was substantially decreased compared to the control (Figure 1B,C).

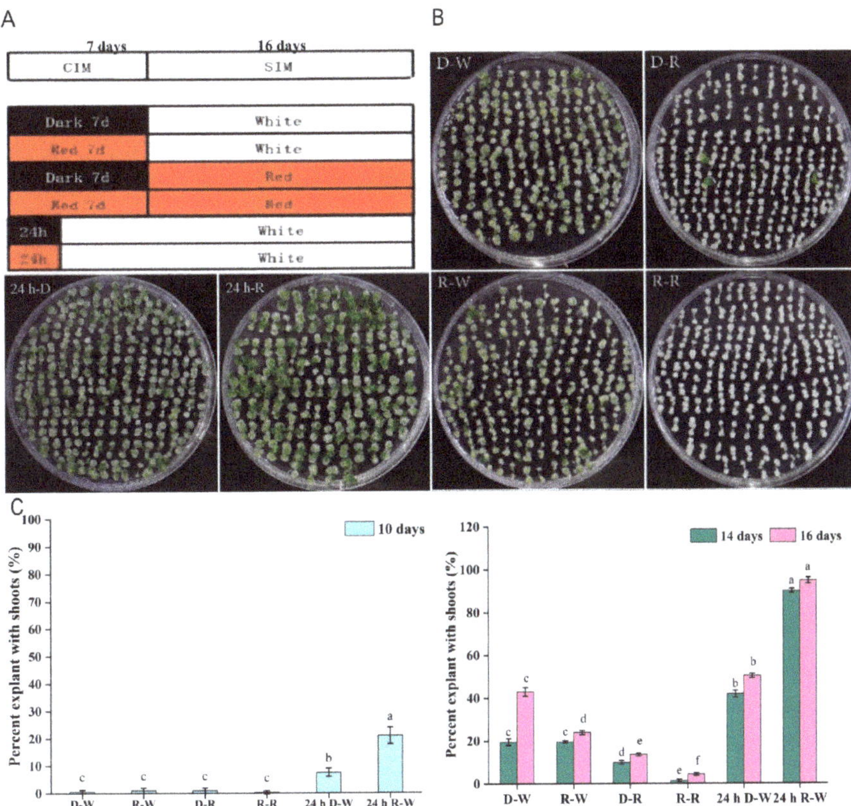

Figure 1. Effects of treatments with combinations of darkness, red, and white light on shoot regeneration in *Arabidopsis* Col-0 callus. (**A**) Light-combinations used in the callus-induction medium (CIM) and shoot-induction medium (SIM) stages. (**B**) Phenotypes of shoots induced under different light treatments at 16 days on the SIM. (**C**) Percent of explants with shoots at 10, 14, and 16 days on the SIM. D-W (the control treatment), dark in the CIM, white light in the SIM. R-W, red light in the CIM and white light in the SIM. D-R, dark in the CIM and red light in the SIM. R-R, red light in both the CIM and the SIM. 24 h D-W, early 24 h dark and then shifting to 6 days white light in the CIM followed by white light throughout the SIM. 24 h R-W, early 24 h red light shifting to 6 days white light in the CIM, followed by white light treatment in the SIM. The above experiments were performed with three biological replicates, each containing 120 root segments of *Arabidopsis thaliana*. Standard errors were calculated from three sets of biological replicates. A significant difference in the percent of explants with shoots between different treatments was analyzed at 10 days, 14 days and 16 days, respectively. The least significant difference method (LSD) was used for significance test ($p < 0.05$); different lowercase letters represent statistical differences in pairwise comparisons between LSD test groups ($p < 0.05$).

To investigate the effects of red light on shoot regeneration during the SIM stage, callus was subjected to dark culture on CIM for 7 days followed by red light treatment in the SIM stage for 16 days (D-R). This treatment resulted in severe inhibition of shoot regeneration at 14 days and 16 days, suggesting that long-term red light treatment in the SIM stage was detrimental to shoot formation. As expected, continuous red-light treatment in CIM and SIM stages (R-R) blocked the capacity for shoot regeneration (Figure 1B,C), demonstrating that long-term exposure to red light weakens the ability to form shoots.

While long-term red light treatment produced negative effects on shoot regeneration regardless of the CIM or SIM stage, short-term light treatment showed contrasting effects. In this study, the light regimens of either 24 h dark (24 h D-W) or 24 h red light (24 h R-W) in the initial stage after root excision and then shifting to white light in the CIM and SIM was used to regulate the shoot generation. These treatments both induced shoot formation, whereas roughly 7.5% and 21% of explants regenerated shoots at 10 days in SIM following 24 h D-W or 24 h R-W, respectively (Figure 1C). No shoot formation was observed at 10 days in the SIM stage if either 7 days-culture in darkness or 7 days of red light treatment was used in the CIM stage (Figure 1C), indicating that the initial 24 h of treatment under dark or red light was sufficient for shoot induction. Subjection to the 24 h R-W treatment after root excision significantly increased the percentage of explants with shoots, where about 90% and 94.8% of explants regenerated shoots at 14 days and 16 days. Similarly, 24 h D-W treatment after root excision also promoted shoots regeneration, where about 41.8% and 50.3% of explants regenerated shoots at 14 days and 16 days (Figure 1C). In contrast, long term exposure to darkness or red light in the CIM stage decreased the potential for shoot generation, where about 20% and less than 40% of explants regenerated shoots at 14 days and 16 days (Figure 1C).

2.2. Effects of Different Light Regimens on Shoot Growth Vigor, Distribution Pattern and Shoot Regeneration Number per Explant

Based on the above results, darkness or red light during the initial 24 h after root excision significantly increased the percentage of explants with shoots, while the effect on the shoot number per explant, shoot growth vigor and shoot distribution pattern was not studied. Here, 120 individual explants in each replicate were collected, and the average shoot number per explant was calculated. As shown in Figure 2A,C, the shoot morphology and distribution pattern were distinctly different between these treatments. Specifically, red light treatment in the CIM and white light treatment in the SIM (R-W) significantly inhibited shoot growth, resulting in the development of small and abnormal shoots (Figure 2B). In contrast, regenerated shoot number and size increased following the control D-W, supporting that continuous darkness not continuous red light treatment in CIM facilitated the shoot development. The 24 h R-W and 24 h D-W treatment conditions both significantly increased the shoot generation frequency, how about the effect on shoot number per explant? Compared with the control D-W treatment, the 24 h R-W and 24 h D-W treatments presented the different effects on shoot number per explant, where the average shoot number per explant was slightly increased under 24 h D-W but decreased under 24 h R-W. Notably, 24 h R-W greatly improved shoot growth vigor and changed the shoot distribution patterns. Unlike the weak growth vigor and the wide distribution pattern of shoots under D-W, R-W and 24 h D-W conditions, most shoots generated after 24 h R-W treatment developed into seedling with multiple normal leaves, and emerged from the middle location of the callus after the 24 h R-W treatment (Figure 2A,C).

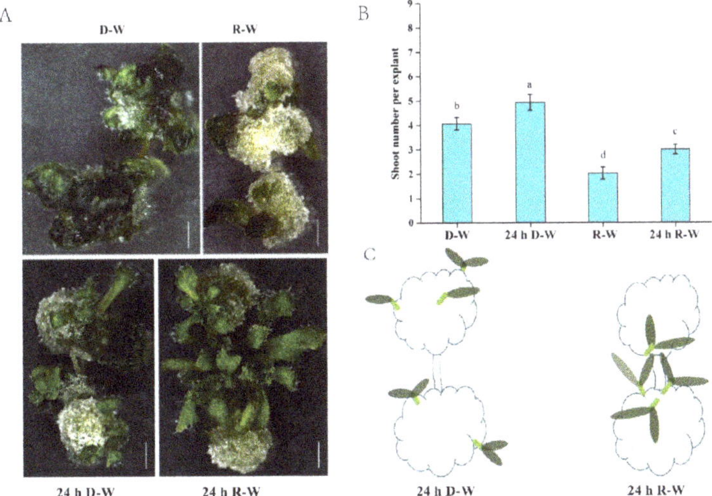

Figure 2. Effects of different light regimens on shoot regeneration number per explant, shoot growth vigor and shoot distribution. (**A**) Shoot morphology of explants under different treatments. The shoot growth vigor was severely inhibited under the R-W treatment, but was promoted under the 24 h R-W treatment. (**B**) Shoot number per explant under different treatments. In contrast with other treatments, average shoot number per explant under R-W treatment was significantly decreased. (**C**) Shoot distribution pattern under different treatments. The shoots were centralized to the middle location under the 24 h R-W treatment, but were widely distributed around callus under the 24 h D-W treatment. The shoot distribution pattern under the D-W or R-W treatments was similar to that of the 24 h D-W treatment. Error bars indicated the standard deviation from three independent experiments, each containing 120 root segments of *Arabidopsis thaliana*. The least significant difference method (LSD) was used for the significance test ($p < 0.05$); different lowercase letters represent statistical differences in pairwise comparisons between LSD test groups ($p < 0.05$).

2.3. NPA Treatment Disrupts the Red Light Induced Shoot Distribution Pattern and Changes the Shoot Number per Explant

Shoot generation was closely associated with the *WUS* expression location and auxin distribution, *WUS* is expressed in the region of low auxin level, and high auxin levels were around the area of *WUS* expression [31]. The guiding hypothesis of this work is that 24 h R-W may regulate the pattern of shoot distribution by controlling the *WUS* location depending on the polarity of auxin distribution. To test this hypothesis, auxin transport inhibitor 1-N-naphthylphtalamic acid (NPA) was added to the CIM at three different concentrations during the early treatment of 24 h red light. Root explants were subsequently transferred to the CIM without NPA for callus induction under white light. Compared with the 24 h red light treatment lacking NPA, the addition of NPA substantially changed shoot numbers and patterns of shoot distribution. Shoot numbers at 16 d on the SIM averaged 3.4 shoots for each explant under 24 h R-W treatment, while the average shoot number per explant increased to 4.4, 5.6, and 7.8 shoots per explant with the addition of 12.5 μM NPA, 25 μM NPA, 50 μM NPA, respectively (Figure 3C). The shoot distribution patterns were also disrupted, the generated shoots were centralized to the middle location of callus under the 24 h R-W treatment, while the shoots were widely distributed around the callus after NPA treatment (Figure 3A,B). A likely cause of this altered phenotype is that NPA interferes with the transport and distribution of auxin that is otherwise regulated by 24 h R-W treatment, the widely spread auxin gradients may facilitate the distribution of the *WUS* signal, thus leading to wider patterns of shoot distribution and an increased average number of shoots.

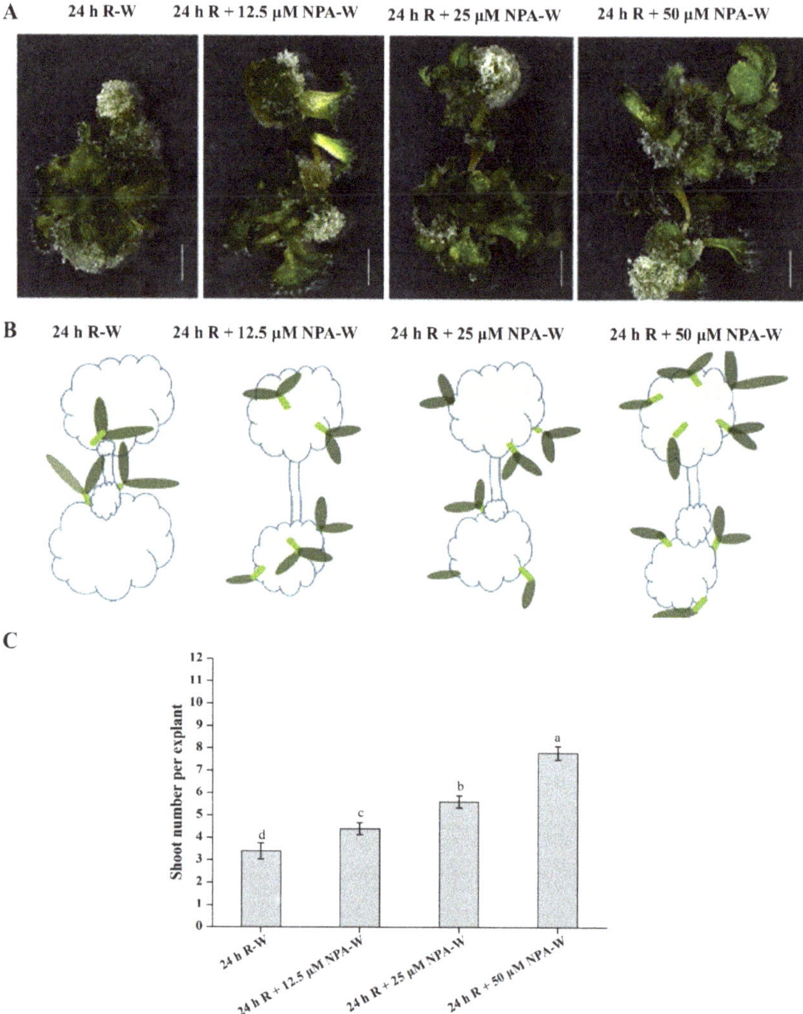

Figure 3. Effects of 1-N-naphthylphtalamic acid (NPA) treatment on average shoot number per explant, the percent explant with shoots and shoot distribution pattern. (**A**) Shoot number and distribution pattern under different treatments. Compared with the shoot morphology under the 24 h R-W treatment, different concentrations of NPA both increased the average shoot number and distributed the shoot distribution pattern. (**B**) Model for shoot distribution pattern under different treatments. Shoots were centralized to the middle location under the 24 h R-W treatment, NPA treatments caused the wide distribution of shoots. (**C**) Effects of NPA concentration on average shoot number per explant. With the increase in NPA concentration, the average shoot number per explant was also increased. The 24 h R-W treatment refers to early 24 h red light shifting to 6 days white light in the CIM, followed by white light treatment in the SIM. The 24 h R + 12.5 µM NPA-W, 24 h R + 25 µM NPA-W, or 24 h R + 50 µM NPA-W treatments refer to: 24 h of red light treatment on CIM containing 12.5, 25, or 50 µM NPA after root excision, then transfer to white light in the CIM and SIM. Error bars indicate the standard deviation from three independent experiments, each containing 120 root segments of *Arabidopsis thaliana*. The least significant difference method (LSD) was used for the significance test ($p < 0.05$); Different lowercase letters represent statistical differences in pairwise comparisons between LSD test groups ($p < 0.05$).

2.4. Dynamic Distribution of WUS under Different Light Regimens and NPA Treatment

The capacity for shoot regeneration is controlled by the level of *WUS* expression, while the location of *WUS* expression determines where and when shoots merge from the callus [14,32]. For the purpose of detecting the location of WUS expression, and thus shoot distribution under different treatments, a pWUS::*WUS*-GUS marker line was used. As shown in Figure 2, the average shoot number per explant after 24 h D-W treatment slightly increased over that of the standard D-W treatment, which is a phenomenon closely associated with *WUS* signal strength and distribution patterns. Specifically, *WUS* signal strength and distribution area across callus cells after the 24 h D-W treatment was slightly increased in comparison with callus cells subjected to the D-W treatment (Figure 4A,B). Expectedly, the weak *WUS* signal under R-W treatment caused a reduction in shoot number and inhibition of shoot growth, whereas the centralized strong *WUS* signal observed under the 24 h R-W treatment promoted shoot growth vigor and the centralized distribution of shoots (Figure 4A,B).

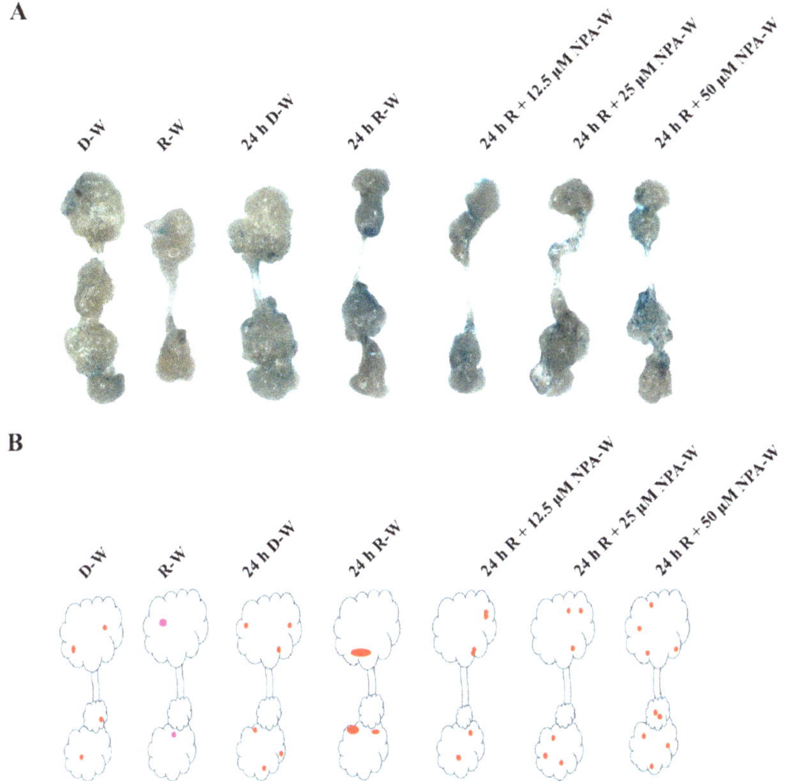

Figure 4. Dynamic localization and expression patterns of *WUS* under different light regimens and concentrations of NPA. (**A**) *WUS* localization patterns at day 7 on the SIM after different treatments. In contrast with D-W, the R-W treatment weakened the *WUS* signal, the 24 h R-W treatment promoted the centralized localization in the middle of the explant, the 24 R-NPA-W treatment increased the *WUS* signal and promoted a wider distribution of *WUS*. (**B**) A model for the *WUS* distribution pattern under different treatments. Red spots indicate strong *WUS* signal, pink spots indicate weak *WUS* signal. The size of spots indicates the area of *WUS* signal. Three biological replicates were performed for each experiment, and each replicate contained 120 root segments. Calluses derived from root segment were stained.

Given that NPA treatment promotes an increase in shoot number per explant following 24 h red light conditions, we further hypothesized that the *WUS* distribution pattern was changed. Similar to the shoot distribution patterns, *WUS* expression was observed to be centralized to specific locations following 24 h red light treatment. NPA treatments significantly disrupted the distribution of *WUS*, and WUS signals were widely expressed in the callus after 7 days on the SIM (Figure 4A,B). Moreover, the WUS signals and distribution area gradually increased commensurately with increased NPA concentration (Figure 4A,B). These results support that WUS expression patterns thus appeared to be regulated by red light, darkness, duration of light treatment, and auxin polar distribution.

2.5. Expression of Marker Genes Involved in Shoot Regeneration and Callus Development Are Dynamically Regulated by Light and NPA

Stem cells within the SAM are necessary during organogenesis and somatic embryogenesis, and in these cells *WUS* gene expression is critical for the regulation of stem cell fate [32]. The pre-incubation stage on the CIM was necessary to activate *WUS* expression to regulate stem cell fate in the SIM stage as described by Shemer et al. [33]. Different to the low expression level of WUS in the CIM, *WUS* expression was significantly induced at 7 days on SIM after the above five treatments, whereas 24 h R-W and 24 h D-W treatments both significantly activated *WUS* expression compared with other treatments (Figure 5A). This finding suggests that high levels of *WUS* expression promoted SAM initiation and shoot formation via regulation of stem cell fate, shown by an increased shoot generation frequency at 10 days, 14 days and 16 days on the SIM after 24 h R-W and 24 h D-W treatments (Figure 1C).

SHOOT MERISTEMLESS (STM) and organ boundary genes CUP SHAPED COTYLEDON1 (*CUC1*), *CUC2*, and *CUC3* regulate each other to establish the embryonic SAM and to specify cotyledon boundaries during embryogenesis [13,34]. Compared with the D-W, R-W and 24 h D-W treatments, the 24 h R-W treatment obviously upregulated *STM*, *CUC1* and *CUC2* expression (Figure 5A), which supports the data showing that the 24 h R-W treatment, after root excision significantly increased the percentage of shoot-bearing explants (up to 94.8%) at 16 days on the SIM (Figure 1B,C). We concluded from these data that shoot formation ability was regulated by the expression level of marker genes depending on which light treatment was applied. Specifically, the higher expression level of *WUS*, *STM*, *CUC1* and *CUC2* under the 24 h R-W treatment compared to other treatments significantly promoted the shoot generation capacity.

WOX5, *PLT3*, *LBD*16 and *LBD*18 are also key genes controlling callus development. The 24 h R-W and 24 h D-W treatments significantly upregulated *WOX5* expression relative to the other five treatments at CIM7 (Figure 5B), suggesting that high levels of *WOX5* expression provided the basis for induction of *WUS* expression, coinciding with the high shoot regeneration rates under the 24 h R-W and 24 h D-W treatments (Figure 1). Compared to CIM 0, *PLT3*, *LBD*16 and *LBD*18 both presented high expression levels at CIM7 under all seven different treatments (Figure 5B), suggesting that *PLT3*, *LBD*16 and *LBD*18 play key roles in mediating the formation of root primordia, which thus provides the foundation for stem cell formation. Primer sequences of marker genes for callus-induction and shoot-induction used in the Tables S4 and S5 during this study.

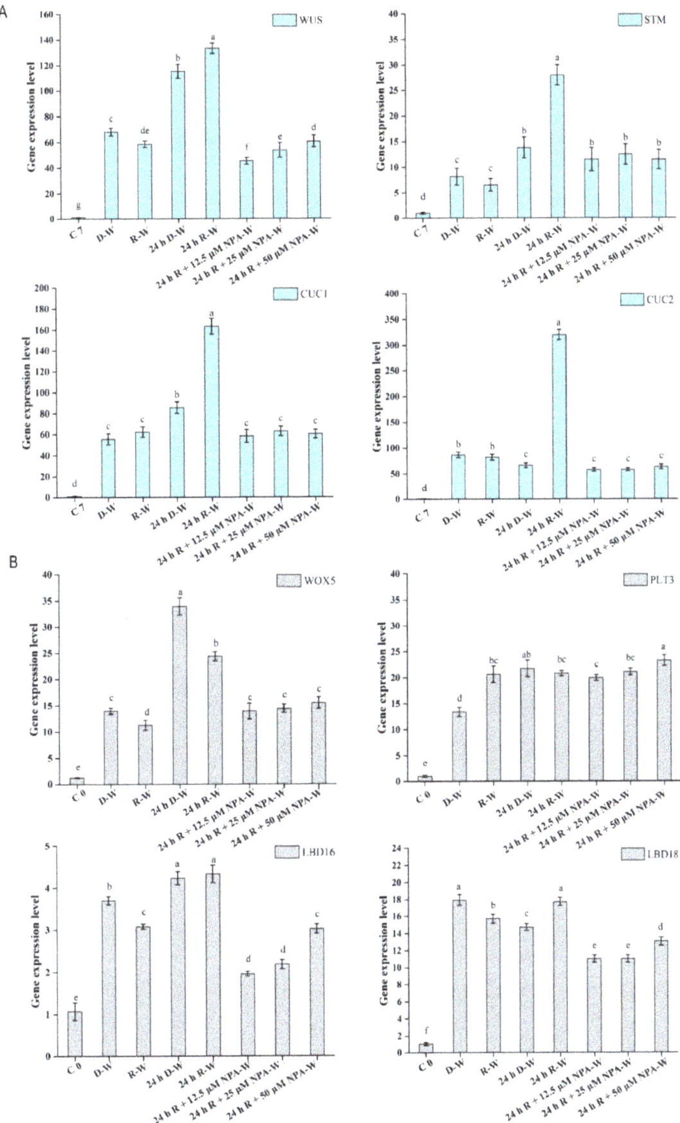

Figure 5. Expression patterns of shoot regeneration marker genes and callus development marker genes under different treatments. (**A**) Expression patterns of marker genes associated with shoot regeneration at day 7 on the CIM and SIM, respectively. (**B**) Expression patterns of marker genes involved in callus development at 7 day on the CIM. C0, root explants on the CIM at day 0, gene expression levels in excised roots at CIM day 0 was set to 1 for quantification of relative expression. C7, root explants on the CIM at day 7, gene expression levels in excised roots at CIM day 7 was set to 1 for quantification of relative expression. Error bars indicate the standard deviation from three independent experiments. A total of 30 individual calluses were collected for qPCR expression analysis for each biological replicate. The least significant difference method (LSD) was used for a significance test ($p < 0.05$); different lowercase letters represent statistical differences in pairwise comparisons between LSD test groups ($p < 0.05$).

2.6. DEGs in CIM and SIM Stages under D-W, 24 D-W and 24 R-W Treatments

Early low-fluence red light or darkness facilitates the shoot regeneration of excised Arabidopsis roots (Figure 1B,C), the samples at CIM0, CIM7, SIM7 under D-W, 24 D-W and 24 R-W treatments were selected to reveal the regulatory mechanism through transcriptome analysis. Analysis of Pearson's correlation coefficient confirmed that the high repeatability among the three biological samples of CIM0, DWCIM7, DCIM7, RCIM7, DWSIM7, DSIM7 and RSIM7 (Supplementary Figure S1A). The three period materials were obviously clustered into three groups including group one (CIM0), group two (DWCIM7, DCIM7, RCIM7) and group three (DWSIM7, DSIM7, RSIM7) (Supplementary Figure S1C). The Venn diagram reflected that 14,482 genes were co-expressed in the CIM stage, but 74, 92 and 195 genes were exclusively expressed in RCIM7, DCIM7 and DWCIM7 (Supplementary Figure S1C), respectively. A total of 15,790 genes were co-expressed in the SIM stage, 226, 259 and 135 genes were exclusively expressed in RSIM7, DSIM7 and DWSIM7, respectively (Supplementary Figure S1D). These results suggest that the shoot regeneration ability was controlled by a large number of common genes and a few private genes.

Differentially expressed genes (DEGs) between different stages and treatments are listed in Supplementary Figure S2A, and DEGs induced by light signal was listed in Supplementary Table S2. We found that auxin-responsive genes, *IAAs* and *ARFs* were expressed in the CIM stage not in the SIM stage (Supplementary Figure S2B). Consistent with the results of quantitatively detected marker genes (Figure 5A,B), *PLTs*, *WOX5*, *WOX11*, *LBD16*, *LBD18*, *LBD19* related with root primordia properties were activated during the CIM stage, but restricted expression in the SIM stage (Supplementary Figures S2C and S3A,B). Besides, expression of auxin efflux carrier *PIN1*, *PIN7* and embryogenesis related genes *BBM*, *AGL15* were obviously induced in the CIM and SIM stages (Supplementary Figure S3C–F). However, *LEC1*, *LEC2*, *ABI3*, *FUS3* related to embryogenesis were still restricted in the CIM stage and in the primary regeneration shoot stage (Supplementary Figure S2D).

2.7. GO and KEGG Enrichment Analysis of DEGs in the CIM and SIM Stage

In order to study the function of DEGs, gene ontology (GO) and the Kyoto Encyclopedia of Genes and Genomes (KEGG) enrichment analysis were performed. Some DEGs were involved in the GO terms, including meristem development, embryo development; plant hormone response, transport, biosynthesis, signal; cellular response to red light, far red light, dark (Supplementary Figure S6A–C, Supplementary Tables S1–S3). For the DEGs between DWCIM7, DWSIM7, RCIM7, RSIM7, DCIM7 and DSIM7, the GO terms "cell differentiation", "maintenance of shoot apical meristem identity", "stem cell population maintenance" and "shoot apical meristem specification" were significantly enriched (Supplementary Table S1). For the DEGs between DWCIM7, RCIM7, DCIM7, the GO terms: "response to red light or far red light", "cellular response to light stimulus" were significantly enriched (Supplementary Table S2). For the DEGs between DWCIM7 and CIM0, GO terms related with plant hormone signaling, including: "response to auxin", "auxin polar transport", "abscisic acid transport" and "response to cytokinin, jasmonic acid and oxygen signal" were significantly enriched (Supplementary Table S3).

KEGG pathway enrichment analysis of DEGs showed that the plant hormone signal pathway, starch and sucrose metabolism and fatty acid elongation were significantly enriched among different treatments. For the transitional process from callus to shoot regeneration, 3637 common DEGs, 1111 and 1085 private DEGs were detected between DCIM7 vs. DSIM7 and RCIM7 vs. RSIM7 (Supplementary Figure S5A), and these DEGs involved in the KEGG pathways: "Plant hormone signal transduction", "Brassinosteroid biosynthesis", "Fatty acid elongation", "Starch and sucrose metabolism" were significantly enriched (Supplementary Figure S5B). For the CIM7 stage (callus dedifferentiation), 22 DEGs between DCIM7 and RCIM7, 607 DEGs between DWCIM7 and DCIM7, 424 DEGs between DWCIM7 and RCIM7 were detected (Supplementary Figure S5C), and these DEGs involved in KEGG pathways, "Phenylpropanoid biosynthesis" and "Indole alkaloid biosynthesis" were significantly enriched (Supplementary Figure S5D). For the SIM7 stage (regeneration shoot), 541 DEGs between DSIM7 and RSIM7, 867 DEGs between DWSIM7 and DSIM7, 266 DEGs between DWSIM7 and RSIM7 were detected (Supplementary Figure S5E), and these DEGs involved in KEGG pathways: "Phenylpropanoid biosynthesis", "Protein processing in endoplasmic reticulum", "alpha-Linolenic acid metabolism", and "Starch and sucrose metabolism" were significantly enriched (Supplementary Figure S5F). Some representative DEGs involved in enriched KEGG pathways are summarized in Table 1 (transitional stage, DCIM7 vs. DSIM7, RCIM7 vs. RSIM7), Table 2 (dedifferentiation stage, DWCIM7 vs. DCIM7, DCIM7 vs. RCIM7, DWCIM7 vs. RCIM7), Table 3 (regeneration shoot, DWSIM7 vs. DSIM7, DWSIM7 vs. RSIM7, DSIM7 vs. RSIM7).

Table 1. Significant representative differentially expressed genes (DEGs) involved in KEGG enrichment during the transitional stage (DCIM7 vs. DSIM7, RCIM7 vs. RSIM7).

Gene ID	Gene Name	Pathway	KO ID	Corrected p-Value	Rich Facter
AT1G02850	*BGLU11*	Phenylpropanoid biosynthesis	ko00940	6.15×10^{-8}	1.96
AT1G15820	*CP24*	Photosynthesis—antenna proteins	ko00196	1.00×10^{-5}	3.41
AT1G03130	*PSAD2*	Photosynthesis	ko00195	3.84×10^{-5}	2.18
AT1G02850	*BGLU11*	Starch and sucrose metabolism	ko00500	4.14×10^{-3}	1.53
AT1G01120	*KCS1*	Fatty acid elongation	ko00062	5.64×10^{-3}	2.40
AT2G26710	*BAS1*	Brassinosteroid biosynthesis	ko00905	6.11×10^{-3}	3.92
AT1G04240	*IAA3*	Plant hormone signal transduction	ko04075	1.31×10^{-2}	1.42
AT1G12900	*GAPA2*	Carbon fixation in photosynthetic organisms	ko00710	1.52×10^{-2}	1.89
AT1G09420	*G6PD4*	Carbon metabolism	ko01200	0.02	1.42
AT1G02850	*BGLU11*	Cyanoamino acid metabolism	ko00460	0.03	1.91
AT1G12550	*HPR3*	Glyoxylate and dicarboxylate metabolism	ko00630	0.03	1.80
AT2G19190	*FRK1*	Plant–pathogen interaction	ko04626	3.62	1.51

Corrected p-value ≤ 0.05; RCIM7 (24 h R-W treatment, CIM 7 d); DCIM7 (24 h D-W treatment, CIM 7 d); RSIM7 (24 h R-W treatment, SIM 7 d); DSIM7 (24 h D-W treatment, SIM 7 d); 24 h D-W, early 24 h dark and then shifting to 6 days white light in CIM followed by white light throughout SIM; 24 h R-W, early 24 h red light shifting to 6 days white light in CIM, followed by white light treatment in SIM; CIM, callus induction medium; SIM, shoot induction medium; KO, KEGG Ortholog.

Table 2. Significant representative differentially expressed genes (DEGs) involved in KEGG enrichment during the dedifferentiation stage (DWCIM7 vs. DCIM7, DCIM7 vs. RCIM7, DWCIM7 vs. RCIM7).

Gene ID	Gene Name	Pathway	KO ID	Corrected p-Value	Rich Facter
AT2G40890	*REF8*	Flavonoid biosynthesis	ko00941	0.04	7.14
AT3G44540	*FAR4*	Cutin, suberine and wax biosynthesis	ko00073	4.66×10^{-3}	6.78
AT1G74000	*SS3*	Indole alkaloid biosynthesis	ko00901	1.47×10^{-3}	20.00
AT1G51680	*4CL1*	Phenylalanine metabolism	ko00360	6.07×10^{-6}	7.86
AT1G05260	*RCI3*	Phenylpropanoid biosynthesis	ko00940	0	6.75

Corrected p-value ≤ 0.05; RCIM7 (24 h R-W treatment, CIM 7 d); DCIM7 (24 h D-W treatment, CIM 7 d); DWCIM7 (D-W treatment, CIM 7 d); D-W (the control treatment); 24 h D-W, early 24 h dark and then shifting to 6 days white light in CIM followed by white light throughout SIM; 24 h R-W, early 24 h red light shifting to 6 days white light in CIM, followed by white light treatment in SIM; CIM, callus induction medium; SIM, shoot induction medium; KO, KEGG Ortholog.

Based on the GO and KEGG analysis results, we proposed a possible model for revealing the mechanism controlling the capacity of shoot regeneration and callus formation under the early low-fluence red light or darkness (Figure 6). Table S6 was the overall situation of KEGG enrichment pathway genes corrected in the transition stage (DCIM7 vs. DSIM7, RCIM7 vs. RSIM7). Table S7 was the overall situation of KEGG enrichment pathway genes corrected in the dedifferentiation stage (DWCIM7 vs. DCIM7, DCIM7 vs. RCIM7, DWCIM7 vs. RCIM7). Table S8 was the overall situation of KEGG enrichment pathway genes corrected in the primary regeneration shoot stage (DWSIM7 vs. DSIM7, DWSIM7 vs. RSIM7, DSIM7 vs. RSIM7). In the callus induction (Figure 6A) and shoot regeneration process (Figure 6B), plant hormones transduction including auxin signaling, jasmonate signal transduction, brassinosteroid signal transduction, gibberellic acid mediated signaling, abscisic acid(ABA)-induced signal transduction and cytokinin signal transduction were enriched (Supplementary Table S3). Meanwhile, we found that fatty acid elongation, brassinosteroid biosynthesis, red light signaling, starch and sucrose metabolism, carbon fixation, carbon metabolism and cutin, suberine and wax biosynthesis also participated in the process of shoot regeneration.

Table 3. Significant representative differentially expressed genes (DEGs) involved in KEGG enrichment during the primary regeneration shoot stage (DWSIM7 vs. DSIM7, DWSIM7 vs. RSIM7, DSIM7 vs. RSIM7).

Gene ID	Gene Name	Pathway	KO ID	Corrected p-Value	Rich Facter
AT1G26560	*BGLU40*	Phenylpropanoid biosynthesis	ko00940	2.05×10^{-5}	3.14
AT1G04980	*PDI10*	Protein processing in endoplasmic reticulum	ko04141	3.76×10^{-3}	2.37
AT1G17420	*LOX3*	alpha-Linolenic acid metabolism	ko00592	5.59×10^{-3}	4.89
AT1G06020	*FRK3*	Starch and sucrose metabolism	ko00500	1.94×10^{-2}	2.23
AT1G72450	*JAZ6*	Plant hormone signal transduction	ko04075	0.10	1.89
AT4G28720	*YUC8*	Tryptophan metabolism	ko00380	0.16	3.49
AT1G02920	*GST11*	Glutathione metabolism	ko00480	0.19	2.59

Corrected p-value ≤ 0.05; RSIM7 (24 h R-W treatment, SIM 7 d); DSIM7 (24 h D-W treatment, SIM 7 d) and DWSIM7 (D-W treatment, SIM 7 d); D-W (the control treatment); 24 h D-W, early 24 h dark and then shifting to 6 days white light in CIM followed by white light throughout SIM; 24 h R-W, early 24 h red light shifting to 6 days white light in CIM, followed by white light treatment in SIM; CIM, callus induction medium; SIM, shoot induction medium; KO, KEGG Ortholog.

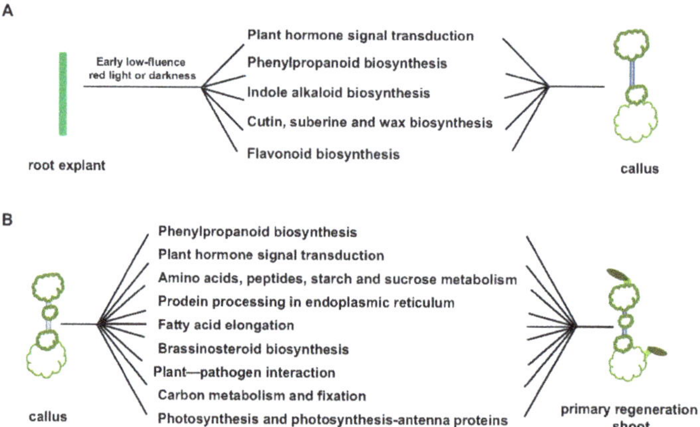

Figure 6. Functional enrichment of differential genes in the CIM and SIM stages under the early low-fluence red light or darkness. (**A**) Differential gene enrichment analysis revealed the biological pathway for the transition from root explant to callus in DCIM7 vs. RCIM7, DWCIM7 vs. DCIM7 and DWCIM7 vs. RCIM7. (**B**) Differential gene enrichment analysis revealed the biological pathway for the transition from callus to primary regeneration shoot in the SIM stages (DWSIM7 vs. DSIM7, DWSIM7 vs. RSIM7 and DSIM7 vs. RSIM7) and in the transitory stage from CIM to SIM (DCIM7 vs. DSIM7 and RCIM7 vs. RSIM7). RCIM7 (24 h R-W treatment, CIM 7 d); DCIM7 (24 h D-W treatment, CIM 7 d); DWCIM7 (D-W treatment, CIM 7 d); RSIM7 (24 h R-W treatment, SIM 7 d); DSIM7 (24 h D-W treatment, SIM 7 d) and DWSIM7 (D-W treatment, SIM 7 d); D-W (the control treatment); 24 h D-W, early 24 h dark and then shifting to 6 days white light in CIM followed by white light throughout SIM; 24 h R-W, early 24 h red light shifting to 6 days white light in CIM, followed by white light treatment in SIM; CIM, callus induction medium; SIM, shoot induction medium.

3. Discussion

3.1. Early Red Light or Dark Exposure on Excised Root Tissue Improved the Shoot Regeneration Capacity via Regulation of WUS Signal Strength and Distribution Pattern

Explants, historically, are often placed in a continuous dark or light treatment immediately after excision in *Arabidopsis* [6]. In terms of monochromatic light treatments, some studies revealed that shoot regeneration was inhibited by blue/UV-A wavelengths [19]. Additionally, low far red (FR)reduced chloroplast xanthophyll pigments, and was not sufficient to elicit ROS, leading to inhibition of shoot regeneration [35].

In this study, our data suggest that root explants may be highly susceptible to conditions in the initial 24 h following excision. We found a significant benefit was gained from exposure to darkness or low-fluence red light treatment during the first 24 h after root excision, leading to an increase in the shoot regeneration frequency over that of callus exposed to continuous darkness in the CIM stage (Figure 1B,C). These results support data showing that darkness or low-fluence red light exposure during the initial 24 h after root excision regulates long-term shoot regeneration in *Arabidopsis* ecotype Columbia. Low-fluence red light increased the biosynthesis and transport of free indole-3-acetic acid (IAA) in the *Arabidopsis* meristem, cotyledons, hook and hypocotyl, and also promoted auxin biosynthesis in cucumber seedlings [29,30,36]. We found that the accumulation of auxin in the dark treatment for 3 days (D-3d) was lower than that after 24 h of red light treatment and continued white light treatment until 3 days (R-3d) (Supplementary Figure S4a,c). Similar results were observed at 5 days between D-5d and R-5d (Supplementary Figure S4b,d). Low-fluence red light may promote shoot regeneration because of the accumulation of auxin in the early stage. The capacity for shoot generation reported to be closely related with the endogenous auxin gradient [1]. So, we hypothesized that the initial treatment with 24 h of low-fluence red light may change the auxin distribution gradient via regulating auxin biosynthesis and transport, and finally promote shoot generation frequency. To test this hypothesis, NPA treatments were used to block the polar transport of endogenous auxin, average shoot number per explant was increased along with the changed auxin gradient. A likely cause of this altered shoot number was that altered auxin polar gradient regulated the *WUS* polar distribution.

The induction of *WUS* is the most critical event in the shoot formation phase, which is controlled by interaction of auxin and cytokinin [37]. Auxin-induced *WUS* expression is essential for embryonic stem cell renewal during somatic embryogenesis and de novo shoot regeneration in *Arabidopsis* [12,31]. Here we show that *WUS* distribution pattern and signal strength were both changed by the initial 24 h red light or 24 h darkness treatment in comparison with the control D-W treatment. In this case, *WUS* was localized in the middle location of the callus, which lead to a centralized distribution of shoots and increased shoot growth vigor. Different from the 24 h red light treatment, the 24 h darkness treatment increased the *WUS* signal distribution area, which increased the average shoot number per explant and changed the shoot distribution pattern. We believe that early 24 h red light or 24 h darkness treatments both modulated the polarity of auxin distribution but caused the different auxin distribution pattern. The auxin distribution pattern induced by the 24 h red light treatment led to centralized expression and localization of *WUS*.

3.2. Low-Fluence Red Light Increased the Capacity for Shoot Regeneration Depending on Upregulation of WOX5, LBD16, LBD18, PLT3, WUS, STM, CUC1, and CUC2

LBD16 and *LBD18* may function redundantly in the establishment of a root primordium-like identity in the newly formed callus. Induction of *LBD16* on the CIM was found to be necessary to gain pluripotency in the callus, which thus modulated the ability for shoot generation, while inhibition of *LBD16* expression blocked the capacity for shoot development in the callus [8]. Higher expression of *LBD16* promoted the gain of callus pluripotency, resulting in formation of root founder cells. Subsequent activation of *WOX5* and *PLT3* synergistically, was previously reported to promote the fate transition from root founder cells to root primordium cells [11]. In this study, low-fluence red light upregulated *WOX5, LBD16, LBD18,* and *PLT3* expression, thus promoting the gain of callus pluripotency and root primordium formation.

The initial treatment condition of 24 h of red light after root excision significantly upregulated the expression levels of *WUS, STM, CUC1,* and *CUC2,* relative to their expression following the other treatments at 7 days on the SIM (Figure 5), suggesting that the simultaneous high expression of these four genes synergistically increased the capability for shoot regeneration. The *WUS* gene is critical for regulation of stem cell fate in plants, and low-fluence red light first induces high *WUS* expression during the SIM stage. This high expression of *WUS* specifies stem cell fate to promote the initiation of the SAM. Subsequently, expression of *CUC1* and *CUC2* are functionally redundant in the induction of SAM formation, through activation of the *STM.* Similar studies have also shown that overexpression of *CUC1* and *CUC2* genes in *Arabidopsis* promoted adventitious shoot formation on callus tissue via activation of *STM* expression [34], while *GhWUS* from *Gossypium hirsutum* promoted de novo shoot regeneration in *Arabidopsis* calluses by directly activating *CLV3* and *CUC2* [14]. Taken together, these data show that low-fluence red light promoted the expression of callus development genes *WOX5, PLT3, LBD16,* and *LBD18,* and also activated the shoot generation marker genes *WUS, STM, CUC1,* and *CUC2,* leading to a capacity for high shoot regeneration.

Our results indicate that the initial 24 h of treatment under dark or red light was sufficient for shoot induction. Treatment with NPA increased the average shoot number and caused wider distribution of shoots on calluses, a likely cause of this phenotype was that the dynamic distribution pattern of *WUS* expression was disrupted by the endogenous auxin gradient. However, the regulatory mechanism of *WUS* expression and dynamic distribution by red light remain to be elucidated. Increasing evidence suggests that red light affects auxin synthesis and transport [29,30]. Thus, it is possible that red light regulates *WUS* expression level and location depending on the auxin polar distribution, and the regulatory network needs to be further established.

3.3. GO and KEGG Enrichment Analysis of DEGs Discover the Vital Regulatory Pathway Underlying Early Low-Fluence Red Light or Darkness

GO enrichment analysis found that DEGs are mainly involved in meristem development, cell differentiation, response to red light or far red light, response to auxin, and auxin polar transport. KEGG enrichment analysis showed that plant hormone signal transduction, carbon metabolism, starch and sucrose metabolism, fatty acid elongation, brassinosteroid biosynthesis pathways were significantly enriched. Table S9 was FPKM values of all genes. And the Table S10: All databases for gene annotation. These significantly enriched pathways and GO terms mainly included auxin response and transport genes (*IAAs, PINs,* and *ARFs*), meristem development genes (*WOX5, PLT3, LBD16, WOX*11), fatty acid elongation genes (*KCSs*), brassinosteroid biosynthesis genes (*BAS1*), suggesting that these above DEGs regulate shoot generation capacity controlled by early low-fluence red light or darkness. The callus initiation and gain of pluripotency is regulated by a number of transcription factors such as *WOX5, WOX*11, *WOX*12 and *LBDs* [9,11,38,39]. Very-long-chain fatty acids (*KCS1*) restrict regeneration capacity by confining pericycle competence for callus formation in *Arabidopsis* [40].

We also predicted other types of transcription factors, ERF, AP2, RSK, ARF, BES1, BSD, BUB, IAA and so on (Supplementary Figure S6d), these DEGs provided potential genes for establishing the regulatory network of shoot development, and their differential expression may play an important role in *Arabidopsis* shoot regeneration. Our research was the tip of the iceberg regarding the results of transcriptome analysis. Therefore, the mechanism controlling the capacity of shoot regeneration and callus formation under the early low-fluence red light or darkness needs further study.

4. Materials and Methods

4.1. Plant Materials

The *Arabidopsis thaliana* plants used in this study were of the wild-type Columbia (Col-0) genetic background. The pDR5::GUS::GFP line, which reflects the auxin level by monitoring auxin responsiveness, was used for Western blot analyses. The pWUS::WUS-GUS marker line was kindly provided by Professor Lin Xu (Institute of Plant Physiology and Ecology, China Academy of Science, China).

4.2. Plant Growth and In Vitro Culture

The *A. thaliana* seeds were surface-sterilized in tubes and then spread on seed germination Murashige and Skoog (MS) medium (1× MS salts, 2% sucrose, 0.3% Gelrite gellan gum, pH 5.7). The plates were kept at 4 °C in darkness for 48 h to overcome seed dormancy, after which they were placed in a greenhouse at 20–22 °C under a 16 h light/8 h dark photoperiod for two weeks. To collect the excised roots as explants, seedlings at 14 d post-germination were cut and root tissue was subsequently cultured on solid callus induction medium (CIM; 1× Gamborg's B5 salts, 3% sucrose, 0.5 g/L MES, 0.05 mg/L kinetin, 0.5 mg/L 2,4-D, and 0.3% Gelrite gellan gum, pH 5.7) under different light combinations for 7 days at a constant temperature of 22 °C. After cultivation for 7 days, the explants were transferred into the shoot induction medium (SIM; 1× MS salts, 1% sucrose, 0.5 mg/L MES, 2 mg/L zeatin, 1 mg/L d-biotin, 0.4 mg/L IBA, 0.3% Gelrite gellan gum, pH 5.7) under continuous light at 22 °C. The shoots on each explant were defined as being at least 1 mm long. Shoot regeneration frequency was obtained by measuring the rate of the number of explants with shoot derived from total number of explants cultured on SIM. Average shoot number per explant were calculated by measuring the rate of the total number of shoots derived from the number of explants with shoots. The shoot regeneration frequency under different light combination treatments was calculated at 10, 14 and 16 days and the shoot number per explant was calculated at 16 days after transfer to the SIM medium. The experiments were performed with three biological replicates, each containing 120 root segments of *Arabidopsis thaliana*.

4.3. Lighting Conditions

Twenty-four hours of darkness, continuous high white light (photosynthetic photon flux density (PPFD): 80–90 µmol m^{-2} s^{-1}) and constant temperature were provided by a cold light source plant growth box, which is required for the induction of callus. An LED plant growth lamp was used to induce shoots and provided a suitable intensity of red light (PPFD: 40–60 µmol m^{-2} s^{-1}). Different light combinations were used during the culture process (Figure 1A). Low-fluence red light treatment means continuous red light for 24 h photoperiod, darkness treatment means continuous dark for 24 h photoperiod, white light treatment means continuous white light for 24 h photoperiod. D-W (the control treatment), dark in CIM, white light in SIM. R-W, low-fluence red light in CIM and white light in SIM. D-R, dark in CIM and low-fluence red light in SIM. R-R, low-fluence red light in both the CIM and SIM. The 24 h D-W treatment involved early 24 h dark and then shifting to 6 days white light in the CIM followed by white light throughout the SIM. While the 24 h R-W treatment involved early 24 h low-fluence red light shifting to 6 days white light in the CIM, followed by white light treatment in the SIM. After different combinations of light treatment, samples at CIM 0 d, CIM 7 d (SIM 0 d), and SIM 7 d were collected for qRT-PCR analysis.

4.4. Western Blot Analyses

The pDR5::GUS::GFP transgenic plants were used to reveal GFP protein expression levels. GFP-fusion transgenic plants were used for Western blot analyses with anti-GFP antibodies. The ratio of gray values, which reflects the relative expression of protein, was equal to the ratio of the gray value of the GFP protein to the gray value of the internal control. The ACTIN protein was used as an internal control. The software ImageJ was used to calculate the gray value.

4.5. Total RNA Isolation and Quantitative Real-Time (qRT)-PCR Analysis

Total RNAs was isolated from the samples collected. The PrimeScriptTMRT reagent Kit with gDNA Eraser (TaKaRa) was used to remove genomic DNA from the total RNA and to obtain cDNA. The sequences of all the qRT-PCR primers are provided in Supplementary Tables S1 and S2. The cDNA was diluted four to five times and then used as a template for the qRT-PCR. For the qRT-PCR, actin2 (AtACT2, AT3G18780) was used as an internal standard, and the gene expression level of CIM day 0 was set to 1 for quantification of relative expression. Three biological replicates were carried out for this experiment, and 30 individual calluses as a biological replicate were collected for qPCR expression analysis.

4.6. RNA-Seq Analysis

After different combinations of light treatment, samples at CIM0 (CIM 0 d), RCIM7 (24 h R-W treatment, CIM 7 d), DCIM7 (24 h D-W treatment, CIM 7 d), DWCIM7 (D-W treatment, CIM 7 d), RSIM7 (24 h R-W treatment, SIM 7 d), DSIM7 (24 h D-W treatment, SIM 7 d) and DWSIM7 (D-W treatment, SIM 7 d) were collected for RNA-seq. Genes with a log2 fold change $\geq$ 2 were classified as being significantly up-regulated/down-regulated in samples at CIM0, RCIM7, DCIM7, DWCIM7, RSIM7, DSIM7 and DWSIM7. RNA-Seq expression was standardized as fragments per kilobase million (FPKM).

4.7. Chemical Inhibitor

Different concentrations of 1-N-naphthylphtalamic acid (NPA) were used as follows: 12.5 µM, 25 µM, 50 µM 1-N-naphthylphtalamic acid (NPA) and dimethyl sulfoxide (DMSO). The NPA was dissolved in DMSO, which was filter-sterilized and added to CIM after autoclaving. The roots of 14-day-old seedlings (post germination) were excised and placed on CIM containing NPA for 24 h of culture under low-fluence red light. Seedlings were then transferred to CIM without NPA for 6 days under white light and finally transferred to SIM for shoot induction. The experiments were performed with three biological replicates, each containing 100 to 120 root segments of *Arabidopsis thaliana*.

4.8. β-GUS Assay

GUS chemical tissue staining experiments were performed according to protocols in a previous study [8]. To clearly observe the GUS staining, the stained tissues were decolorized with an alcohol concentration gradient. Under the stereo microscope, the gene expression level and localization of expression were observed in specific tissues via GUS staining. In this study, the pWUS::WUS-GUS marker lines were used to perform tissue staining after 7 days of induction on SIM. The above experiments were performed with three biological replicates, each containing 120 root segments, callus derived from root segment were stained.

5. Conclusions

The results of this work revealed that early low-fluence red light or darkness promotes the shoot regeneration capacity of excised *Arabidopsis* roots. NPA treatment disrupts the red light induced shoot distribution pattern and changes dynamic distribution of *WUS*. The 24 h D-W and 24 h R-W treatments obviously upregulated expression of marker genes involved in shoot regeneration and callus development, such as *WUS*, *STM*, *CUC1*, *WOX5* and *LBD*16. GO and KEGG enrichment analysis found that DEGs are mainly involved in meristem development, cell differentiation, response to red light or far red light, response to auxin, auxin polar transport and plant hormone signal transduction, carbon metabolism, starch and sucrose metabolism, fatty acid elongation, brassinosteroid biosynthesis pathways. The findings of this study provided fundamental evidence into the mechanism of shoot regeneration, which will support future functional examination of vital molecular mechanisms of shoot regeneration.

Supplementary Materials: The following are available online at http://www.mdpi.com/2223-7747/9/10/1378/s1, Figure S1: Overall qualitative analysis of the transcriptomics data. Figure S2: Number of DEGs and heat maps of auxin responsive and meristem development genes at comparison of different processing combinations. Figure S3: Transcriptome analysis of marker genes expression patterns in the CIM and SIM stages. Figure S4: A Western blot shows the kinetic of auxin accumulation in D-3d, R-3d, D-5d and R-5d. Figure S5: Differential genes commonly and KEGG pathway enrichment analysis of DEGs. Figure S6: Gene ontology (GO) annotation and transcription factor prediction in the three stages. Table S1: The meristem development genes were found according to GO analysis. Table S2: The genes of response to red light, far red light and dark were found according to GO analysis. Table S3: The genes of plant hormone response, transport, biosynthesis and oxygen signal were found according to GO analysis. Table S4: Primer sequences of marker genes for callus-induction used in this study. Table S5: Primer sequences of marker genes for shoot-induction used in this study. Table S6: The overall situation of KEGG enrichment pathway genes corrected in the transition stage (DCIM7 vs. DSIM7, RCIM7 vs. RSIM7). Table S7: The overall situation of KEGG enrichment pathway genes corrected in the dedifferentiation stage (DWCIM7 vs. DCIM7, DCIM7 vs. RCIM7, DWCIM7 vs. RCIM7). Table S8: The overall situation of KEGG enrichment pathway genes corrected in the primary regeneration shoot stage (DWSIM7 vs. DSIM7, DWSIM7 vs. RSIM7, DSIM7 vs. RSIM7). Table S9: FPKM values of all genes. Table S10: All databases for gene annotation.

Author Contributions: X.G. and F.L. conceived the original screening and research plan. Y.D. and X.W. performed the experiments. X.W., Y.W. provided technical assistance to Y.D. and analyzed the data. X.G. and Y.D. wrote the manuscript. All authors have read and agreed to the published version of the manuscript.

Funding: This research was funded by the National Natural Science Foundation of China (31701476).

Conflicts of Interest: The authors declare no conflict of interest.

Abbreviations

AGL15	Agamous-like 15
ARF	Auxin response factor
BBM	Baby boom
CUC2	Cup-shaped cotyledon 2
CUC1	Cup-shaped cotyledon 1
CLV3	Clavata 3
CIM0	CIM 0 d
CIM	Callus induction medium
D-W	the control treatment
DWCIM7	D-W treatment, CIM 7 d
DWSIM7	D-W treatment, SIM 7 d
DCIM7	24 h D-W treatment, CIM 7 d
DSIM7	24 h D-W treatment, SIM 7 d
DEGs	Differentially expressed genes
D-R	dark in CIM and red light in SIM
GO	Gene Ontology
IAA	Indole-3-acetic acid inducible
JAZ	Jasmonate ZIM-domain
KEGG	Kyoto encyclopedia of genes and genomes
LBD	LOB domain-containing protein 2
NPA	The auxin transport in-hibitor 1-N-naphthylphthalamic acid
PCA	Principal component analysis
PIN	PIN-FORMED
ROS	Reactive oxygen species
RAM	Root apical meristems
RCIM7	24 h R-W treatment, CIM 7 d
RSIM7	24 h R-W treatment, SIM 7 d
R-W	red light in CIM and white light in SIM

R-R	red light in both CIM and SIM
SAM	Shoot apical meristem
SIM	Shoot induction medium
WUS	Wuschel
WOX	WUSCHEL-related homeobox
24 h D-W	early 24 h dark and then shifting to 6 days white light in CIM followed by white light throughout SIM
24 h R-W	early 24 h red light shifting to 6 days white light in CIM, followed by white light treatment in SIM

References

1. Zhang, H.; Zhang, T.T.; Liu, H.; Shi, D.Y.; Zhang, X.S. Thioredoxin-Mediated ROS Homeostasis Explains Natural Variation in Plant Regeneration. *Plant Physiol.* **2018**, *176*, 2231–2250. [CrossRef]
2. Birnbaum, K.D.; Alvarado, A.S. Slicing across kingdoms: Regeneration in plants and animals. *Cell* **2008**, *132*, 697–710. [CrossRef] [PubMed]
3. Xu, L.; Huang, H. Genetic and Epigenetic Controls of Plant Regeneration. *Curr. Top. Dev. Biol.* **2014**, *108*, 1–33. [PubMed]
4. Duclercq, J.M.; Sangwan-Norreel, B.; Catterou, M.; Sangwan, R.S. De novo shoot organogenesis: From art to science. *Trends Plant Ence* **2011**, *16*, 597–606. [CrossRef] [PubMed]
5. Sugimoto, K.; Meyerowitz, E.M. *Regeneration in Arabidopsis Tissue Culture*; Humana Press: Totowa, NJ, USA, 2013.
6. Zhao, Q.H.; Fisher, R.; Auer, C. Developmental phases and STM expression during *Arabidopsis* shoot organogenesis. *Plant Growth Regul.* **2002**, *37*, 223–231. [CrossRef]
7. Che, P.; Lall, S.; Howell, S.H. Developmental steps in acquiring competence for shoot development in *Arabidopsis* tissue culture. *Planta* **2007**, *226*, 1183–1194. [CrossRef]
8. Liu, J.; Hu, X.; Qin, P.; Prasad, K.; Hu, Y.; Xu, L. The WOX11-LBD16 Pathway Promotes Pluripotency Acquisition in Callus Cells During De Novo Shoot Regeneration in Tissue Culture. *Plant Cell Physiol.* **2018**, *59*, 739–748. [CrossRef]
9. Liu, J.; Sheng, L.; Xu, Y.; Li, J.; Yang, Z.; Huang, H.; Xu, L. WOX11 and 12 Are Involved in the First-Step Cell Fate Transition during de Novo Root Organogenesis in Arabidopsis. *Plant Cell* **2014**, *26*, 1081–1093. [CrossRef]
10. Fukaki, H.; Tasaka, M. Hormone interactions during lateral root formation. *Plant Mol. Biol.* **2009**, *69*, 437–449. [CrossRef]
11. Kareem, A.; Durgaprasad, K.; Sugimoto, K.; Du, Y.; Pulianmackal, A.J.; Trivedi, Z.B.; Abhayadev, P.V.; Pinon, V.; Meyerowitz, E.M.; Scheres, B.; et al. PLETHORA Genes Control Regeneration by a Two-Step Mechanism. *Curr. Biol.* **2015**, *25*, 1017–1030. [CrossRef]
12. Cheng, Z.J.; Wang, L.; Sun, W.; Zhang, Y.; Zhou, C.; Su, Y.H.; Li, W.; Sun, T.T.; Zhao, X.Y.; Li, X.G. Pattern of auxin and cytokinin responses for shoot meristem induction results from the regulation of cytokinin biosynthesis by AUXIN RESPONSE FACTOR3. *Plant Physiol.* **2013**, *161*, 240–251. [CrossRef] [PubMed]
13. Ikeuchi, M.; Ogawa, Y.; Iwase, A.; Sugimoto, K. Plant regeneration: Cellular origins and molecular mechanisms. *Development* **2016**, *143*, 1442–1451. [CrossRef] [PubMed]
14. Xiao, Y.; Chen, Y.; Ding, Y.; Wu, J.; Wang, P.; Yu, Y.; Wei, X.; Wang, Y.; Zhang, C.; Li, F. Effects of GhWUS from upland cotton (*Gossypium hirsutum L.*) on somatic embryogenesis and shoot regeneration. *Plant Ence* **2018**, *270*, 157–165. [CrossRef] [PubMed]
15. Meng, W.J.; Cheng, Z.J.; Sang, Y.L.; Zhang, M.M.; Rong, X.F.; Wang, Z.W.; Tang, Y.Y.; Zhang, X.S. Type-B *ARABIDOPSIS* RESPONSE REGULATORs Specify the Shoot Stem Cell Niche by Dual Regulation of WUSCHEL. *Plant Cell* **2017**, *29*, 1357. [CrossRef] [PubMed]
16. Zhang, T.Q.; Lian, H.; Zhou, C.M.; Xu, L.; Jiao, Y.; Wang, J.W. A Two-Step Model for de novo Activation of WUSCHEL during Plant Shoot Regeneration. *Plant Cell* **2017**, *29*, 1073–1087. [CrossRef] [PubMed]
17. Yoshida, S.; Mandel, T.; Kuhlemeier, C. Stem cell activation by light guides plant organogenesis. *Genes Dev.* **2011**, *25*, 1439–1450. [CrossRef] [PubMed]
18. Dong, N.; Montanez, B.; Creelman, R.A.; Cornish, K. Low light and low ammonium are key factors for guayule leaf tissue shoot organogenesis and transformation. *Plant Cell Rep.* **2006**, *25*, 26–34. [CrossRef]

19. Peterman, E.J.; Gradinaru, C.C.; Calkoen, F.; Borst, J.C.; van Grondelle, R.; van Amerongen, H. Xanthophylls in light-harvesting complex II of higher plants: Light harvesting and triplet quenching. *Biochemistry* **1997**, *36*, 12208–12215. [CrossRef]

20. Nameth, B.; Dinka, S.J.; Chatfield, S.P.; Morris, A.; English, J.; Lewis, D.; Oro, R.; Raizada, M.N. The shoot regeneration capacity of excised *Arabidopsis* cotyledons is established during the initial hours after injury and is modulated by a complex genetic network of light signalling. *Plant Cell Environ.* **2012**, *36*, 68–86. [CrossRef]

21. Nishimura, T.; Mori, Y.; Furukawa, T.; Kadota, A.; Koshiba, T. Red light causes a reduction in IAA levels at the apical tip by inhibiting de novo biosynthesis from tryptophan in maize coleoptiles. *Planta* **2006**, *224*, 1427–1435. [CrossRef]

22. Smets, R.; Jie, L.; Prinsen, E.; Verbelen, J.P.; Onckelen, H.A.V. Cytokinin-induced hypocotyl elongation in light-grown *Arabidopsis* plants with inhibited ethylene action or indole-3-acetic acid transport. *Planta* **2005**, *221*, 39–47. [CrossRef] [PubMed]

23. Saitou, T.; Tachikawa, Y.; Kamada, H.; Watanabe, M.; Harada, H. Action spectrum for light-induced formation of adventitious shoots in hairy roots of horseradish. *Planta* **1993**, *189*, 590–592. [CrossRef]

24. Saitou, T.; Hashidume, A.; Tokutomi, S.; Kamada, H. Reduction of phytochrome level and light-induced formation of adventitious shoots by introduction of antisense genes for phytochrome A in horseradish hairy roots. *Plant Cell Tissue Organ Cult.* **2004**, *76*, 45–51. [CrossRef]

25. Saitou, T.; Tokutomi, S.; Harada, H.; Kamada, H. Overexpression of phytochrome A enhances the light-induced formation of adventitious shoots on horseradish hairy roots. *Plant Cell Rep.* **1999**, *18*, 754–758. [CrossRef]

26. Bertram, L.; Lercari, B. Evidence against the involvement of phytochrome in UVB-induced inhibition of stem growth in green tomato plants. *Photosynth. Res.* **2000**, *64*, 107–117. [CrossRef]

27. Ye, S.; Shao, Q.; Xu, M.; Li, S.; Wu, M.; Tan, X.; Su, L. Effects of Light Quality on Morphology, Enzyme Activities, and Bioactive Compound Contents in Anoectochilus roxburghii. *Front. Plant Ence* **2017**, *8*, 857. [CrossRef]

28. Heringer, A.S.; Reis, R.S.; Passamani, L.Z.; de Souza-Filho, G.A.A.; Santa-Catarina, C.; Silveira, V. Comparative proteomics analysis of the effect of combined red and blue lights on sugarcane somatic embryogenesis. *Acta Physiol. Plant* **2017**, *39*, 52. [CrossRef]

29. Liu, X.; Cohen, J.D.; Gardner, G. Low-Fluence Red Light Increases the Transport and Biosynthesis of Auxin. *Plant Physiol.* **2011**, *157*, 891–904. [CrossRef]

30. Rubinstein, B. Auxin and red light in the control of hypocotyl hook opening in beans. *Plant Physiol.* **1971**, *48*, 187–192. [CrossRef] [PubMed]

31. Su, Y.H.; Zhao, X.Y.; Liu, Y.B.; Zhang, C.L.; Zhang, X.S. Auxin-induced WUS expression is essential for embryonic stem cell renewal during somatic embryogenesis in Arabidopsis. *Plant J.* **2010**, *59*, 448–460. [CrossRef]

32. Yadav, R.K.; Perales, M.; Gruel, J.; Girke, T.; Jonsson, H.; Reddy, G.V. WUSCHEL protein movement mediates stem cell homeostasis in the *Arabidopsis* shoot apex. *Genes Dev.* **2012**, *25*, 2025–2030. [CrossRef] [PubMed]

33. Shemer, O.; Landau, U.; Candela, H.; Zemach, A.; Williams, L.E. Competency for shoot regeneration from *Arabidopsis* root explants is regulated by DNA methylation. *Plant Sci. Int. J. Exp. Plant Biol.* **2015**, *238*, 251–261. [CrossRef] [PubMed]

34. Demmig-Adams, B.; Adams III, W.W. The CUP-SHAPED COTYLEDON Genes Promote Adventitious Shoot Formation on Calli. *Plant Cell Physiol.* **2003**, *44*, 113–121.

35. Demmig-Adams, B.; Adams III, W.W. Photoprotection and Other Responses of Plants to High Light Stress. *Annu. Rev. Plant Physiol. Plant Mol. Biol.* **1992**, *43*, 599–626. [CrossRef]

36. Shinkle, J.R.; Kadakia, R.; Jones, A.M. Dim-Red-Light-Induced Increase in Polar Auxin Transport in Cucumber Seedlings. *Plant Physiol.* **1998**, *116*, 1505–1513. [CrossRef]

37. Gordon, S.P.; Heisler, M.G.; Reddy, G.V.; Ohno, C.; Das, P.; Meyerowitz, E.M. Pattern formation during de novo assembly of the *Arabidopsis* shoot meristem. *Development* **2007**, *134*, 3539–3548. [CrossRef]

38. Fan, M.; Xu, C.; Xu, K.; Hu, Y. Lateral Organ Boundaries Domain transcription factors direct callus formation in *Arabidopsis* regeneration. *Cell Res.* **2012**, *22*, 1169. [CrossRef]

39. Iwase, A.; Mitsuda, N.; Koyama, T.; Hiratsu, K.; Kojima, M.; Arai, T.; Inoue, Y.; Seki, M.; Sakakibara, H.; Sugimoto, K. The AP2/ERF Transcription Factor WIND1 Controls Cell Dedifferentiation in Arabidopsis. *Curr. Biol.* **2011**, *21*, 508–514. [CrossRef]

40. Shang, B.; Xu, C.; Zhang, X.; Cao, H.; Xin, W.; Hu, Y. Very-long-chain fatty acids restrict regeneration capacity by confining pericycle competence for callus formation inArabidopsis. *Proc. Natl. Acad. Sci. USA* **2016**, *113*, 5101–5106. [CrossRef]

Publisher's Note: MDPI stays neutral with regard to jurisdictional claims in published maps and institutional affiliations.

MDPI

St. Alban-Anlage 66

4052 Basel

Switzerland

Tel. +41 61 683 77 34

Fax +41 61 302 89 18

www.mdpi.com

Plants Editorial Office

E-mail: plants@mdpi.com

www.mdpi.com/journal/plants